WORKBOOK

AQA A-LEVEL

Geography

PHYSICAL GEOGRAPHY

T0187181

Philip Banks

HODDER
EDUCATION
AN HACHETTE UK COMPANY

Photo credits: Philip Banks

Every effort has been made to trace all copyright holders, but if any have been inadvertently overlooked, the Publishers will be pleased to make the necessary arrangements at the first opportunity.

Although every effort has been made to ensure that website addresses are correct at time of going to press, Hodder Education cannot be held responsible for the content of any website mentioned in this book. It is sometimes possible to find a relocated web page by typing in the address of the home page for a website in the URL window of your browser.

Hachette UK's policy is to use papers that are natural, renewable and recyclable products and made from wood grown in well-managed forests and other controlled sources. The logging and manufacturing processes are expected to conform to the environmental regulations of the country of origin.

Orders: please contact Hachette UK Distribution, Hely Hutchinson Centre, Milton Road, Didcot, Oxfordshire, OX11 7HH. Telephone: +44 (0)1235 827827. Email: education@hachette. co.uk. Lines are open from 9 a.m. to 5 p.m., Monday to Friday. You can also order through our website: www.hoddereducation.co.uk.

ISBN: 978 1 3983 3241 6

First published in 2021 by
Hodder Education,
An Hachette UK Company
Carmelite House
50 Victoria Embankment
London EC4Y 0DZ

www.hoddereducation.co.uk

Impression number   10 9 8 7 6 5 4

Year   2025 2024

Cover photo: Cherries/Adobe Stock

Illustrations by Integra

Typeset by Integra Software Services Pvt. Ltd., Puducherry, India.

Printed by CPI Group (UK) Ltd, Croydon, CR0 4YY

A catalogue record for this title is available from the British Library.

MIX
Paper | Supporting
responsible forestry
FSC™ C104740

Contents

1 Water and carbon cycles 5
- Water and carbon cycles as natural systems
- The water cycle
- The carbon cycle
- Water, carbon, climate and life on Earth

Exam-style questions **16**

2 Hot desert systems and landscapes 20
- Deserts as natural systems
- Systems and processes
- Arid landscape development in contrasting settings
- Desertification

Exam-style questions **35**

3 Coastal systems and landscapes 39
- Coasts as natural systems
- Systems and processes
- Coastal landscape development
- Coastal management

Exam-style questions **56**

4 Glacial systems and landscapes 60
- Glaciers as natural systems
- The nature and distribution of cold environments
- Systems and processes
- Glaciated landscape development
- Human impacts on cold environments

Exam-style questions **81**

5 Hazards 85
- The concept of hazard in a geographical context
- Plate tectonics
- Volcanic hazards
- Seismic hazards
- Storm hazards
- Fires in nature

Exam-style questions **107**

About this book

1   **This workbook** will help you to prepare for Paper 1 of the AQA A-level Geography exam.

2   **Paper 1** is 2 hours 30 minutes long and includes a range of questions. The exam is divided into three sections. These are:
- Section A: Water and carbon cycles (36 marks)
- Section B: either Hot desert systems and landscapes **or** Coastal systems and landscapes **or** Glacial systems and landscapes (36 marks)
- Section C: either Hazards **or** Ecosystems under stress (48 marks)

Note that the Ecosystems under stress option is not covered in this workbook.

3   For each section there are a variety of question types:
- short, point-marked answers
- levels-marked responses
- extended prose

4   Where possible the questions in this workbook are scaffolded — they begin with easier, more accessible questions and then work up to more complex questions. Each section ends with exam-style questions, some of which encourage you to think about links across this and other topics, bringing together all your skills and knowledge.

5   You still need to read your textbook and refer to your revision guides and lesson notes.

6   **Marks and assessment objectives** are indicated for exam-style questions so that you can gauge both the detail and balance of content required in your answers.

The assessment objectives (AOs) are:
- AO1: Demonstrate knowledge and understanding of places, environments, concepts, processes, interactions and change, at a variety of scales.
- AO2: Apply knowledge and understanding in different contexts to interpret, analyse and evaluate geographical information and issues.
- AO3: Use a variety of relevant quantitative, qualitative and fieldwork skills in order to investigate questions and issues, interpret, analyse and evaluate data and evidence, and construct arguments and draw conclusions.

7   **Worked answers** are included throughout the practice questions to help you understand how to gain the most marks.

8   Icons next to the question will help you to identify:

 where geographical **skills** are tested

 where questions require you to make **links** across this and other topics

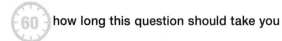 how long this question should take you

9   **Timings** are given for the exam-style questions to make your practice as realistic as possible.

10   Many of the exam-style questions will require more space than is available in this workbook. Answer these on a separate sheet of paper.

11   Answers are available at: **www.hoddereducation.co.uk/workbookanswers**

Topic 1 Water and carbon cycles

Water and carbon cycles as natural systems

A system is a type of geographical model that removes incidental detail to highlight fundamental relationships.

A system is an assemblage of interrelated parts that work together by way of some driving process. It consists of a series of stores or components that have connections between them. There are three types of property: **elements**, **attributes** and **relationships**.

A system has a structure that lies within a boundary. It functions by having inputs and outputs of material (energy and/or matter) that is processed within the components, causing it to change in some way.

The two main types of system used in physical geography are as follows:

• Closed systems, which have transfers of energy both into and beyond the system boundary, but where there is no transfer of matter.

• Open systems, where matter and energy can be transferred from the system across the boundary into the surrounding environment. Most ecosystems, for example, are open systems.

When there is a balance between the inputs and outputs, the system is said to be in a state of dynamic equilibrium. If one of the elements of the system changes, then the stores change and the equilibrium is upset. This is called feedback. There are two types of feedback: positive and negative.

Practice questions ?

1 Draw a labelled sketch of an example of a closed system in physical geography.

2 Complete Figure 1.1, using the following labels:

Component/Store   Output   Input   Flow

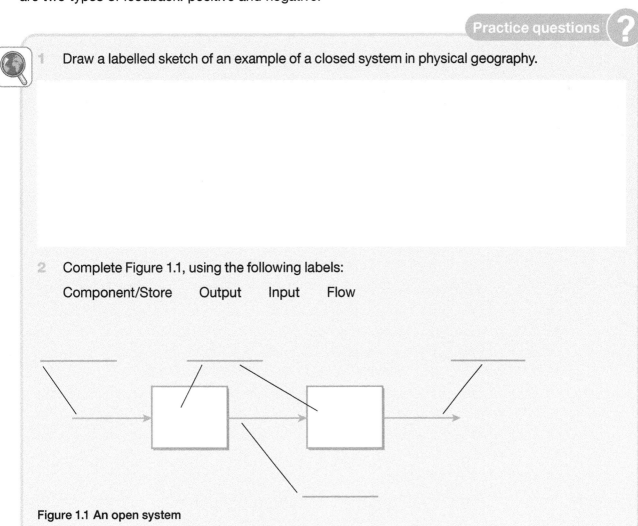

Figure 1.1 An open system

3 Starting with the box labelled 'Global temperature rise', annotate Figure 1.2 with the following labels:

- Carbon dioxide back into the atmosphere
- Warms the oceans
- More carbon dioxide in the atmosphere
- More carbon dioxide to act as a greenhouse gas
- Warm water less able to dissolve gas
- Increased oceanic temperatures
- Dissolved carbon dioxide released by warmer oceans

Give the figure a title.

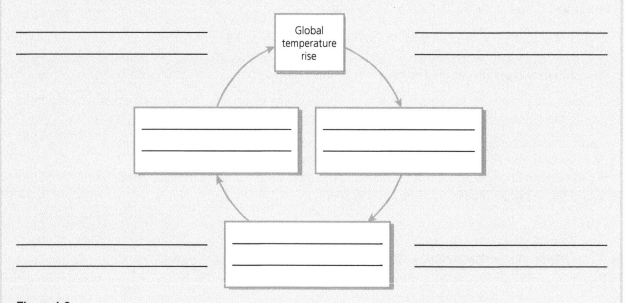

Figure 1.2

4 Complete the following paragraph by filling in the gaps with items from the list below.

Negative feedback

Following a rise in the _____, global carbon dioxide levels

_____. This leads to a global temperature increase which, in

turn, results in _____, meaning that there is an increase in the

_____. This has a _____ and reduces

_____.

List:

take-up of carbon dioxide   increase   dampening effect
use of fossil fuels   increased plant growth   global temperatures

 5 Using an example, explain how a systematic approach can be applied to **one** of the following within a global systems framework:

- Flows of people
- Flows of money
- Flows of ideas and technology

....
....
....
....
....
....
....

The water cycle

Water exists on Earth in three forms: liquid water, solid ice and gaseous water vapour.

A drainage basin (or catchment area) is the area that supplies a river with its water. This includes water found below the water table as well as soil water and any surface flow. A useful way of looking at drainage basins is to consider them as cascading systems. These are a series of open systems that link together so that the output of one is the input of the next.

Within a drainage basin, the balance between inputs and outputs is known as the water balance or budget:

precipitation (P) = discharge (Q) + evapotranspiration (E) ± changes in storage (S)

River levels rise and fall, often showing an annual pattern (called the river's **regime**). They also vary in the short term following heavy rainfall. These short-term changes in river discharge can be displayed using a flood (or storm) hydrograph. Although all storm hydrographs have the same common elements, they are not all the same shape. Their shape is determined by both physical and human factors.

Practice questions

6   Using the data from Table 1.1, describe the distribution of the Earth's water.

Table 1.1

All water	%
Oceanic salt water	97
Fresh water	3

Fresh water	%
Cryospheric water	79
Groundwater	20
Easily accessible surface water	1

Easily accessible surface water	%
Lakes	52
Soil	38
Atmosphere	8
Biomass	1
Rivers	1

....

....

....

....

....

7 Figure 1.3 shows how water moves about a small drainage basin and the nearby ocean. Using Table 1.2, complete Figure 1.3.

Table 1.2

Inputs	Stores	Transfers	Outputs
Precipitation onto land	Lakes and surface water	Overland flow	Evaporation and transpiration from vegetation
Precipitation onto the sea	River channel	Channel flow	Evaporation from water surfaces
	Interception by plants	Infiltration	Runoff from the river
	Soil water	Percolation	Evaporation from the sea
	Groundwater	Throughflow	
		Groundwater flow	
		Throughfall/Stemflow	

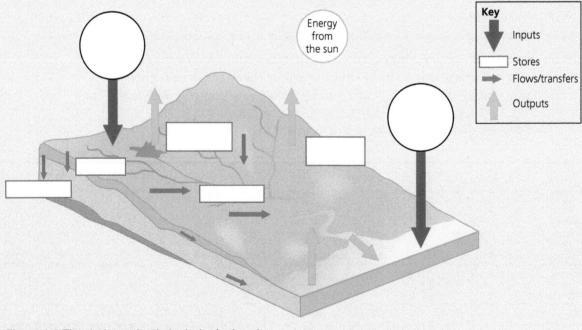

Figure 1.3 The drainage basin hydrological cycle

8 Describe the different ways that water can enter a river channel.

....

....

....

....

....

....

....

....

9　Describe each of the following processes as they apply to the water cycle:

a　Evaporation

....

....

....

....

b　Condensation

....

....

....

....

....

10　Clouds are formed when air is cooled, usually by rising through the atmosphere. Identify the ways in which air can be uplifted and explain the resultant processes that then form clouds and eventually rain.

Worked example

There are three common ways in which air is caused to rise through the atmosphere. First, it can happen when there is a front: where two air masses of different temperatures and densities meet. The warmer, less dense air rises over the colder air. Second, air can be forced to rise over high land. Third, hot land surfaces can heat up local masses of air and cause them to rise by convection.

> Knowledge (AO1): Good summary of the way in which air is forced to rise.

Air expands as it rises. As it does so, it cools down. As the air cools, its ability to hold water vapour is reduced and its relative humidity rises until it is saturated. This is called the dew point temperature and is the point at which condensation begins to occur and cloud droplets are formed.

> Knowledge (AO1): This describes the processes that occur as air rises.

Coalescence of these droplets causes the growth of raindrops.

> Knowledge (AO1): This links the answer to the final part of the question.

11 Complete Figure 1.4 using the following labels:

Bankfull discharge   Rising limb   Storm flow   Peak rainfall   Peak discharge
Base flow   Lag time   Recession limb   Flood water   Rainfall event

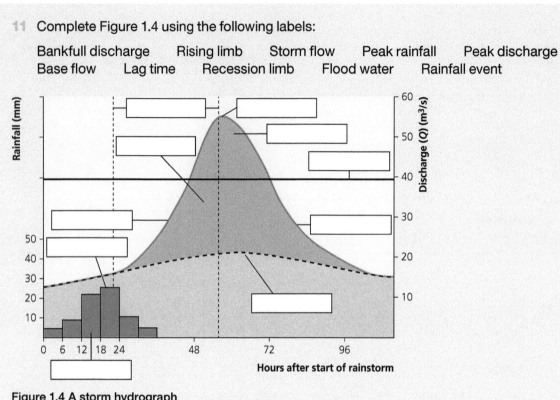

Figure 1.4 A storm hydrograph

12 For **two** named physical factors, outline how they might change the shape of a storm hydrograph.

Physical factor 1 ....

....

....

....

Physical factor 2 ....

....

....

....

13 For **two** named human factors, outline how they might change the shape of a storm hydrograph.

Human factor 1 ....

....

....

....

Human factor 2 ....

....

....

....

....

14 Analyse the seasonal flow of the River Avon, Bath, as shown in Figure 1.5.

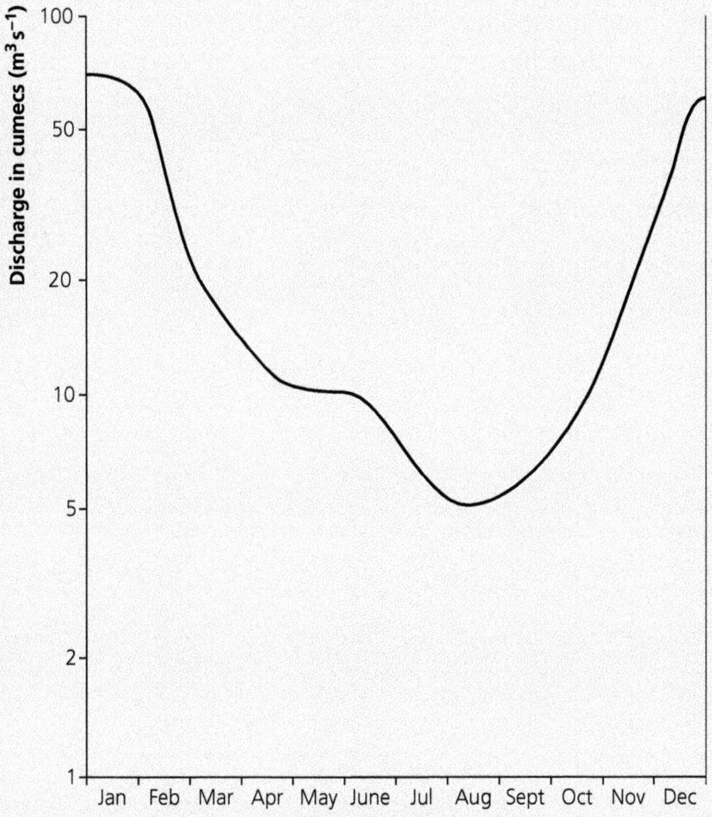

Figure 1.5 Annual flow of the River Avon, Bath

....

....

....

....

....

....

The carbon cycle

Carbon forms more compounds than any other element and scientists believe that there are more than 10 million different carbon compounds in existence today on Earth. It is found in all life forms in addition to sedimentary rocks, diamonds, graphite, coal and petroleum (oil and natural gas).

The global carbon cycle is the pathway by which carbon moves through the Earth system, including the land, oceans, atmosphere and biosphere. Some components of the Earth system, such as the oceans and land, at times act as stores of carbon by storing it for long periods, and at other times act as carbon sources by releasing it back into the atmosphere.

Of growing importance in the global carbon cycle are the emissions from burning hydrocarbons. These are shown in Table 1.3 on page 14.

15 Describe the chemical make-up, occurrence, and importance to the carbon cycle of each of the following:

a Carbon dioxide (CO_2)

....

....

....

b Calcium carbonate ($CaCO_3$)

....

....

....

c Liquid petroleum

....

....

....

16 Describe the following processes. For each one, suggest how it fits into the carbon cycle.

a Photosynthesis

....

....

....

....

....

b   Respiration

....

....

....

....

c   Decomposition

....

....

....

....

....

17 Define the term 'weathering'. Outline the role it plays in the carbon cycle.

....

....

....

....

....

....

....

18 Complete the graph in Figure 1.6 from the data in Table 1.3. You must label the axes and plot
    the data.

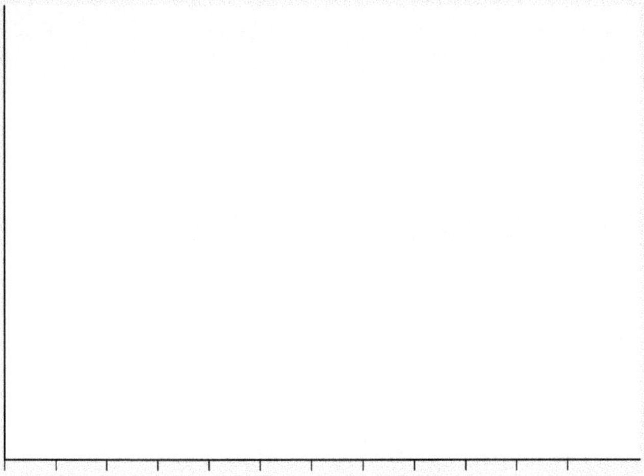

Figure 1.6

Table 1.3 Global carbon emissions from fossil fuels, 1900–2010

Year	Global carbon emissions in millions of metric tonnes of carbon
1900	600
1910	850
1920	1,000
1930	1,100
1940	1,300
1950	1,600
1960	2,600
1970	4,050
1980	5,300
1990	6,100
2000	6,850
2010	9,200

19 Describe the changes in the amount of global carbon emissions as shown in Figure 1.6. What effect could this have on the Earth's climate?

Worked example

In the 110 years between 1900 and 2010, global carbon emissions rose continually so that by the end of the period they were just over 15 times what they were at the beginning. The rise was slow up until 1940 and then they rose rapidly, only slowing down slightly between 1980 and 2000. The most rapid growth of 2,350 mmt was between 2000 and 2010.

Skill (AO3): General trend of changes identified from graph.

Skill (AO3): Accurate use of the graph for more precise description.

The impact of this on global climate could be serious. Carbon emissions are mainly in the form of greenhouse gases. Their build-up in the atmosphere means that they trap outgoing heat radiation from the Earth and cause an atmospheric temperature increase. This is spread around the Earth by winds, tropical storms and ocean currents.

Skill (AO3): Use of the data to draw conclusions. In this case, there is a link between the cause and the effects.

This influences the global pattern of climate. Polar areas will warm up. Trade winds will weaken. Tropical storms will get scarcer, but more intense and longer-lasting. Equatorial areas may become drier while tropical areas will experience longer periods of drought.

Skill (AO3): Some aspects of global climate change identified.

Water, carbon, climate and life on Earth

Without water and carbon, life on Earth would not be possible. Every drop of water cycles continuously through air, land and sea, being used constantly by components within the cycle. Carbon bonds with oxygen to form complex organic compounds that contribute to the existence of living matter.

In the long term, the balance in the amount of carbon in the main global stores (the lithosphere, hydrosphere, biosphere and atmosphere) remains more or less the same. Recent, rapid changes in atmospheric carbon are affecting not only the climate but also the amount of carbon available for plants.

Concerns over the effects of rising atmospheric carbon have led to attempts to reduce or prevent carbon dioxide emissions. These include:

- the development of industrial and power generation processes with reduced emission of greenhouse gases (e.g. solar power generation)

- attempts to capture and store any emissions that do occur, preventing carbon dioxide from returning to the atmosphere (carbon capture and storage, or CCS)

These are collectively known as climate change mitigation.

Practice questions

20 Using an example (e.g. the UK), describe one national strategy to reduce carbon emissions in the power generation industry. Assess the success of this policy.

....

....

....

....

....

....

....

....

....

21 Assess the extent to which **one** of the following human interventions in the carbon cycle has been successful in the mitigation of the impacts of climate change:

Carbon capture and storage **or** Changing agricultural practices **or** Improved aviation practices

....

....

....

....

....

....

....

....

....

Exam-style set 1

1 Outline **two** ways in which water flows through a drainage basin. (AO1)   **4 marks**

....

....

....

....

....

....

 2 Table 1.4 shows data produced by the US National Oceanic and Atmospheric Administration (NOAA). Complete the table and calculate the Spearman's rank correlation coefficient (R_s) using the formula provided. Interpret your answer using the information provided. (AO3)   **6 marks**

Null hypothesis: There is no relationship between global temperature change and average atmospheric carbon dioxide.

Table 1.4 A comparison between global temperatures and average atmospheric carbon dioxide

Year	Average atmospheric temperature compared with long-term average temperature [1900–2010]/°C	Rank R_1	Average CO_2 measured at Mauna Loa Observatory, Hawaii/ppm	Rank R_2	Difference in ranks, d	d^2
1970	0.1	10	328	11	−1.0	1.0
1975	0	11	330	10	1.0	1.0
1980	0.3	8	338	9	−1.0	1.0
1985	0.2	9	346	8	1.0	1.0
1990	0.45	6	353	7	−1.0	1.0
1995	0.5		361			
2000	0.4	7	369	5	2.0	4.0
2005	0.7	3.5	379	4	−0.5	0.25
2010	0.7	3.5	390	3	0.5	0.25
2015	0.9	2	401	2	0	0
2019	1.0	1	415	1	0	0

$$R_s = 1 - \left(\frac{6 \sum d^2}{n^3 - n} \right)$$

where R_s is the Spearman's rank correlation coefficient, n is the number of variables and d is the difference in the ranks of the variables.

For 11 pairs of variables, the critical values for R_s are ± 0.536 at the 5% significance level and ± 0.755 at the 1% significance level.

$\sum d^2 = $ _____ $R_s = $ _____

....

....

....

....

3 Using a completed version of Figure 1.2 (page 6) and your own knowledge, explain how changes in the carbon cycle over time may have an impact on global climate. (AO1, AO2)

(7)

6 marks

....

....

....

....

....

....

....

....

4 As part of your course you have studied a drainage basin at a local level. Assess the impact of precipitation upon the water stores and transfers in that drainage basin and evaluate the extent to which it affects both a sustainable water supply and flooding. (AO1, AO2)

(25)

20 marks

Plan and write your answer on a separate sheet of paper and keep it with your workbook.

Exam-style set 2

(5)

1 Explain the concept of negative feedback in relation to the carbon cycle. (AO1)

4 marks

....

....

....

....

....

....

....

....

2   Study Figure 1.7. Analyse the varying potential for global carbon sequestration from different types of forestry management, 1995–2050. (AO3)

6 marks

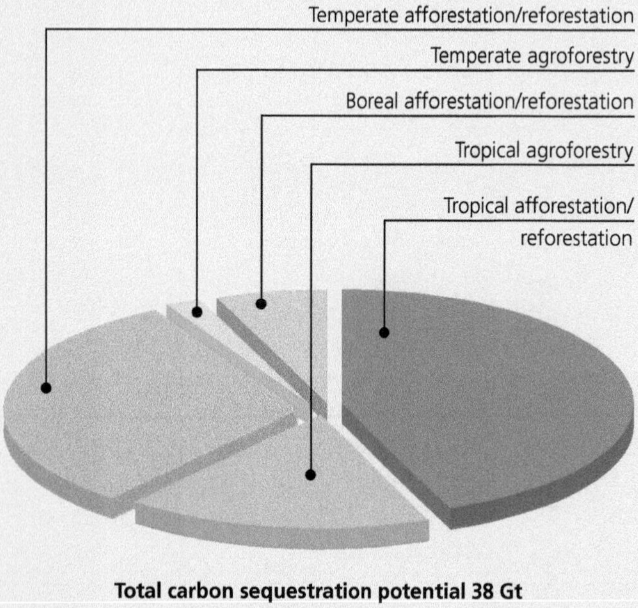

Temperate afforestation/reforestation
Temperate agroforestry
Boreal afforestation/reforestation
Tropical agroforestry
Tropical afforestation/
reforestation

Total carbon sequestration potential 38 Gt

Figure 1.7 Potential contribution of afforestation/reforestation and agroforestry activities to global carbon sequestration, 1995–2050

....

....

....

....

....

....

....

....

....

3   Figure 1.8 shows the predicted change in global rainfall intensity by the end of the twenty-first century. Using Figure 1.8 and your own knowledge, assess the extent to which this change is evenly spread globally. (AO1, AO2)

6 marks

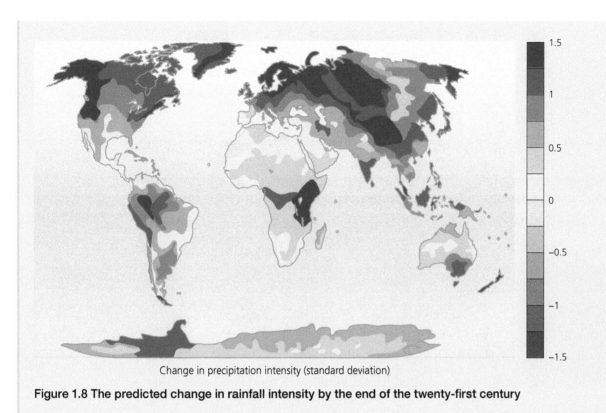

Change in precipitation intensity (standard deviation)

Figure 1.8 The predicted change in rainfall intensity by the end of the twenty-first century

....

....

....

....

....

....

....

....

4   'Human intervention is now the only way to mitigate the impacts of climate change.'

Using examples, assess the extent to which you agree with this
statement. (AO1, AO2)   **20 marks**

Plan and write your answer on a separate sheet of paper and keep it with your workbook.

Additional extended prose questions

Plan and write your answers on separate sheets of paper and keep them with your workbook.

1   Assess the extent to which the emergence of laws and institutions designed to regulate
global systems has influenced carbon transfers and helped to mitigate the impacts of
climate change. (AO1, AO2)   **20 marks**

2   'Natural variations over time have a greater impact than human activity on changes in the
water cycle.'

To what extent do you agree with this statement? (AO1, AO2)   **20 marks**

Topic 2 Hot desert systems and landscapes

Deserts as natural systems

Hot deserts are a form of **dryland**. Drylands can be defined as areas with low precipitation and high evaporation resulting in limited soil moisture. Hot desert areas are found mainly between 15° and 30° north and south of the equator and in the centre of continents or on the west coasts of continents.

Desert systems have inputs, processes, flows and outputs, as shown in Figure 2.1.

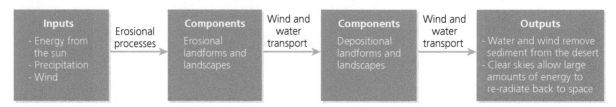

Figure 2.1 A hot desert landscape as an open system

These inputs, processes, flows and outputs act together to form landscapes that include:

- bare rocky plains
- extensive sand seas
- badlands topography
- deep canyons

Soils in hot deserts are coarse-textured, shallow, rocky or gravelly, with little organic matter. There is intense evaporation of water which brings dissolved salts to the surface, leading to saltpans. Desert plants can be categorised as ephemeral, xerophytic, phreatophytic and halophytic.

The water balance compares the mean annual precipitation (P) received with the mean annual potential evapotranspiration (PET). Deserts can be classified using an aridity index (AI). The United Nations has defined the aridity index as:

$$AI = \frac{P}{PET}$$

Hot deserts and their margins are considered to be areas described as hyper-arid to arid, as defined in Table 2.1.

Table 2.1

Classification	AI	Global land area/%
Hyper-arid	AI < 0.05	7.5
Arid	0.05 < AI < 0.20	12.1

There are five main causes of aridity in hot deserts, which can act alone or in combination: atmospheric processes relating to pressure, winds, continentality, relief and cold ocean currents.

1 Describe the global distribution of the main areas of hot desert shown in Figure 2.2.

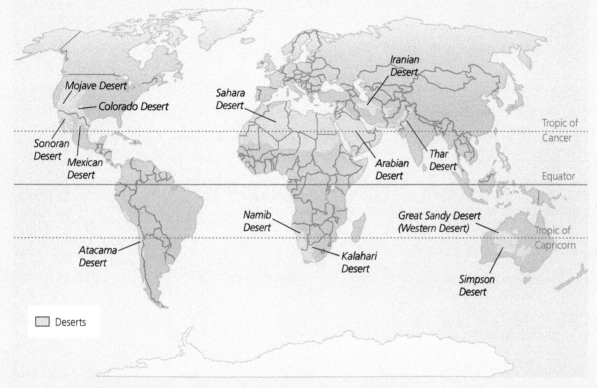

Figure 2.2 Hot desert areas

....

....

....

....

....

....

....

2 Table 2.2 gives details about the climate of Sabha.

Table 2.2 The climate of Sabha

Month	Jan	Feb	Mar	Apr	May	June	July	Aug	Sept	Oct	Nov	Dec
Average maximum temperature/°C	19	21	26	32	36	39	39	39	38	33	26	20
Average minimum temperature/°C	6	8	12	17	22	25	25	25	24	19	12	7
Average precipitation/mm	7	0.5	9	7	1	0.5	0.5	0	0	0	2	1

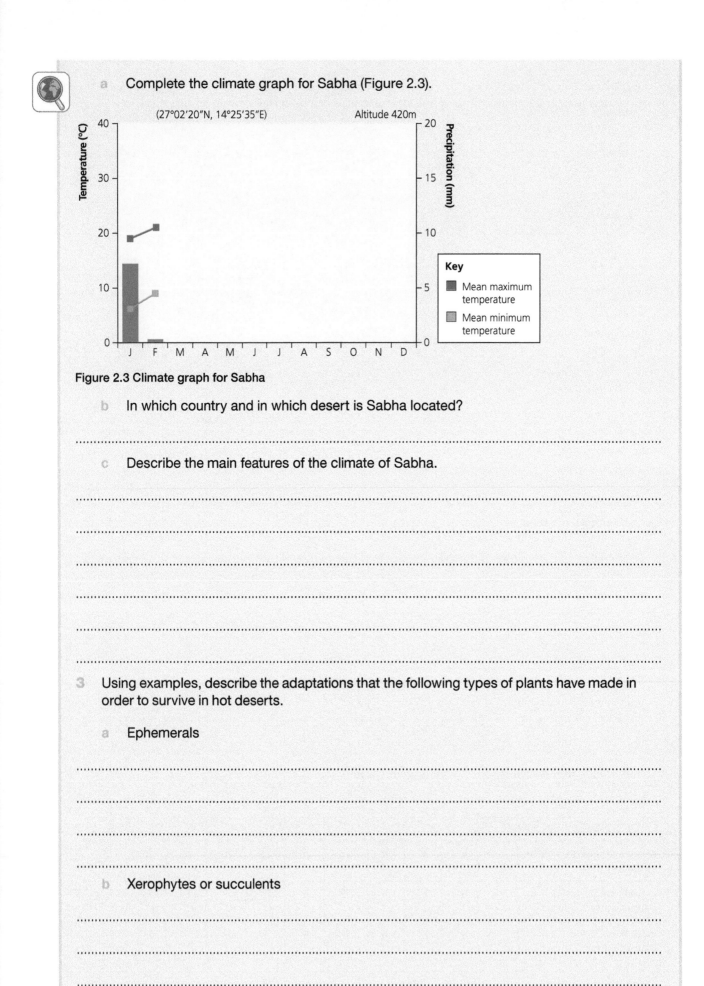

a Complete the climate graph for Sabha (Figure 2.3).

Figure 2.3 Climate graph for Sabha

b In which country and in which desert is Sabha located?

....

c Describe the main features of the climate of Sabha.

....

....

....

....

....

3 Using examples, describe the adaptations that the following types of plants have made in order to survive in hot deserts.

a Ephemerals

....

....

....

b Xerophytes or succulents

....

....

....

c   Phreatophytes

....

....

....

d   Halophytes

....

....

....

4   For each of the following, explain how it contributes to aridity in hot desert areas. For each cause, name a desert affected by it.

a   Subsiding air and associated high atmospheric pressure linked to the descending limbs of the Hadley cell

....

....

....

....

....

b   The rainshadow effect

....

....

....

c   Continentality

....

....

....

....

    d   Cold ocean currents

....

....

....

....

....

5   Explain why there is a wide diurnal temperature range in hot deserts.

....

....

....

....

....

....

....

....

....

....

....

Systems and processes

The sources of energy that drive the desert system are insolation, wind, and water movement (runoff).

Weathering occurs in hot deserts. Chemical weathering is limited by the small amounts of water available in hot deserts but mechanical weathering in the form of exfoliation, thermal fracture, and block and granular disintegration occurs more often.

Wind is almost always present in deserts. Winds are able to pick up (erode) and transport material (which can then itself be a tool in erosion). When their speed falls, they deposit any load they are carrying.

Winds can transport desert material. The method of transport and the distance the particles are carried depend on the strength of the wind and the diameter of the particles. Fine clay and silt particles are lifted high in the air and carried in suspension. Sand and coarse silt are carried by the process of saltation at a height of usually no more than 2m. Coarser material is rolled along the surface by traction.

Water plays (or has played) a major role in shaping desert landscapes. Present-day water comes from:

- localised and rare, but heavy, rainstorms
- exogenous rivers
- endoreic rivers

In wetter periods of the past, water played a major role in shaping the landscape.

6 Outline how two named weathering processes operate in hot deserts.

Process 1 ....

....

....

....

Process 2 ....

....

....

....

....

7 Define the term 'mass movement' and outline its relationship to weathering.

....

....

....

....

....

8 Explain how each of the named processes below contributes to the shaping of desert landscapes. Name one landform shaped predominantly by each process.

a Deflation

....

....

....

....

....

b   Abrasion (sand-blasting)

....

....

....

....

9   Label Figure 2.4 with the following:
    • The heaviest material is rolled along the desert floor by traction (creep)
    • Lighter material bounces along the desert floor (saltation)
    • When a saltated particle falls, it can dislodge further particles
    • The lightest material (dust) is carried by suspension

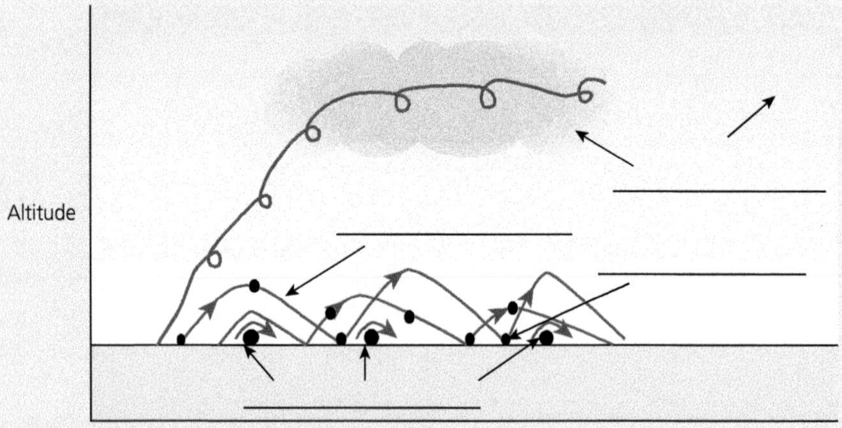

Altitude

Figure 2.4 Sediment transport in deserts

10 There are three main sources of surface water in hot deserts. For each of the following, explain the term and provide an example.

    a   Exogenous river

....

....

....

    b   Endoreic river

....

....

....

    c   Ephemeral river

....

....

....

11 In the context of deserts, outline the differences between channel flash flooding and sheet flooding.

Worked example

In general, flash flooding occurs when there has been a rainstorm of such intensity that the infiltration capacity of the surface is rapidly overcome, and excess water flows downslope. In desert uplands the water becomes confined ●————— into steep-sided, flat-floored gullies called wadis. Because it is confined, the water flows down the wadis with huge velocity and energy. The front of the flood is often described as a 'wall of water'. It can move massive boulders as well as fine debris. The flood dies as rapidly as it developed. ●————

> Knowledge (A01): Clear definition and location of flash flooding in deserts.

> Knowledge (A01): Good description of the nature of flash floods.

Sheet flooding occurs when a number of flash floods, confined in wadis, emerge onto a gently sloping plain at the foot of the hills. The floods spread out and coalesce in an unconfined way and are slowed down by friction. This means coarser material is dropped in alluvial fans close to the base of the hills. Finer material flows down the plain and is deposited as the water either evaporates or soaks into the desert surface. ●————

> Application of knowledge (A02): A link is made between the two to show comparison.

> Knowledge (A01): Good description of the nature of sheet floods.

Arid landscape development in contrasting settings

Although all the above processes are active for at least some of the time, there are landscapes where one set of processes dominates the others. This results in three main desert landscapes:

- **Landscapes and landforms shaped predominantly by wind erosion.** Large areas in deserts consist of flat, stone-covered plains. The wind continually removes fine-grained material, leaving behind deflation hollows, desert pavements, yardangs, zeugen and ventifacts.
- **Landscapes and landforms shaped predominantly by deposition** resulting in accumulations of windblown sand piled up in mounds or ridges. These are known as dunes. There are many types of dune found in deserts, including barchan dunes and seif dunes.
- **Landscapes and landforms shaped predominantly by water.** These include wadis, bajadas, pediments, playas and inselbergs. A wadi is a dry valley that has been formed by an ephemeral stream.

Many deserts are the remains of ancient landscapes. Past processes, acting over a long time (sometimes in different climatic conditions), resulted in the broad landscape. These **residual** or **relict** landscapes have been further shaped by present-day processes.

12 Describe a desert pavement landscape.

....

....

....

....

13 Explain the role of wind in the formation of desert pavements.

....

....

....

....

....

....

14 Explain the role of wind erosion in the formation of a named type of ventifact.

....

....

....

....

....

....

....

15 Yardangs and zeugen depend on the geological structure of rocky desert areas.

Draw a labelled sketch of each type of landform, highlighting their differences. Explain how their different shapes are related to the geological structure.

Worked example

Yardangs

Dominant wind direction

Softer layers of rock are picked out by abrasion to form rock ridges and gullies parallel to the wind

Application (AO2): Impact of geologic structure on how wind abrades the rocks.

Some undercutting occurs

Knowledge (AO1): Geology of the two features shown.

Hard layers of rock are steeply dipping or vertical

Range in size from barely discernible linear steps in the surface to tens of metres high and many kilometres in length

Knowledge (AO1): Clear description of the features.

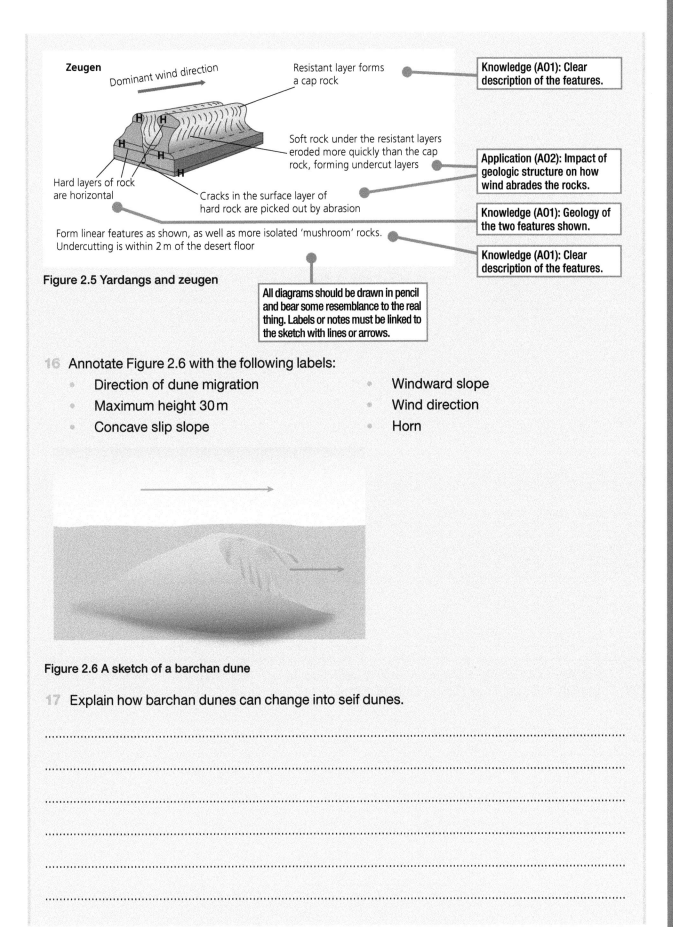

Zeugen

Dominant wind direction →

Resistant layer forms a cap rock

Soft rock under the resistant layers eroded more quickly than the cap rock, forming undercut layers

Hard layers of rock are horizontal

Cracks in the surface layer of hard rock are picked out by abrasion

Form linear features as shown, as well as more isolated 'mushroom' rocks. Undercutting is within 2 m of the desert floor

| Knowledge (A01): Clear description of the features. |

| Application (A02): Impact of geologic structure on how wind abrades the rocks. |

| Knowledge (A01): Geology of the two features shown. |

| Knowledge (A01): Clear description of the features. |

Figure 2.5 Yardangs and zeugen

| All diagrams should be drawn in pencil and bear some resemblance to the real thing. Labels or notes must be linked to the sketch with lines or arrows. |

16 Annotate Figure 2.6 with the following labels:

- Direction of dune migration
- Maximum height 30 m
- Concave slip slope
- Windward slope
- Wind direction
- Horn

Figure 2.6 A sketch of a barchan dune

17 Explain how barchan dunes can change into seif dunes.

....

....

....

....

....

....

18 Outline the main features of wadis and explain the role of episodic water in their formation.

....

....

....

....

....

19 Study Figure 2.7. Using the diagram, explain how bajadas are formed.

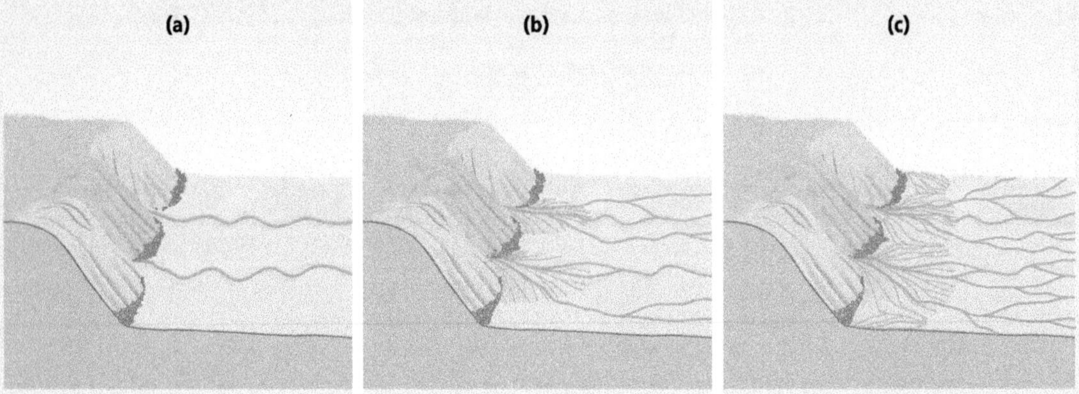

| (a) | (b) | (c) |

Figure 2.7 The formation of a bajada

....

....

....

....

20 Playas or salt lakes are the result of endoreic drainage. Describe a named playa and assess the role that water has played in its formation.

....

....

....

....

....

....

21 Mesas and buttes are forms of inselberg. Using Figure 2.8, assess the extent to which water plays a role in the formation of mesas and buttes.

Figure 2.8 A butte in Arizona, USA

....

....

....

....

....

....

....

....

Desertification

The extent and distribution of hot deserts is not static. Since the maximum of the last glacial advance 18,000 years ago, the overall area of desert lands diminished to a fraction of what they are today before growing again in the last 8,000 years.

The United Nations Convention to Combat Desertification defines the term 'desertification' as 'land degradation in arid, semi-arid and dry sub-humid areas resulting from various factors including climatic variations and human activities'.

Desertification mainly occurs on the dry fringes of the deserts proper. It affects terrestrial areas (topsoil, earth, groundwater reserves and surface runoff), animal and plant populations, and also human activity (population growth and structure, economic activity, settlements and infrastructure). The two main causes of desertification are climate change and human activity.

A recent study of how climate change affects arid regions concluded that:

- deserts are more likely to have increased rainfall

- higher temperatures will also increase evaporation and cause more catastrophic weather events

- weather extremes such as wildfires, cold snaps, heatwaves and storms would intensify

22 Describe the global pattern of those areas at risk of desertification shown in Figure 2.9.

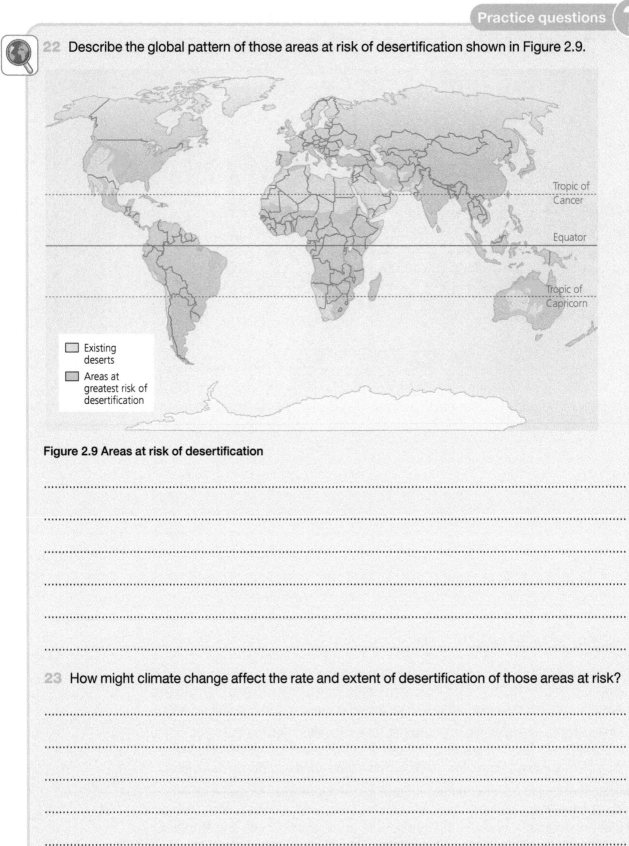

Existing deserts

Areas at greatest risk of desertification

Figure 2.9 Areas at risk of desertification

....

....

....

....

....

....

23 How might climate change affect the rate and extent of desertification of those areas at risk?

....

....

....

....

....

....

....

....

24 Assess the impact of human activity on a named local area undergoing desertification.

....

....

....

....

....

....

....

....

....

....

25 Study Table 2.3. What conclusions can be drawn from these data?

Table 2.3 Estimates of land area subject to desertification belonging to vulnerability classes and corresponding number of impacted population

Vulnerability class	Area subject to desertification		Population affected	
	Area/million km^2	Global land area/%	Number/millions	Global population/%
Low	14.60	11.2	1,085	18.9
Moderate	13.61	10.5	915	15.9
High	7.12	5.5	393	6.8
Very high	7.91	6.1	255	4.4
TOTAL	44.24	34.0	2,648	44.0

....

....

....

....

....

....

....

26 For a named local area where desertification has occurred, describe and explain the impact it has had on the following.

  a The ecosystems found there

....

....

....

....

....

....

   b   The people who live there

....

....

....

....

....

....

....

27 Describe and evaluate the following human responses to desertification in a named local area.
   a   Resilience

....

....

....

....

   b   Mitigation

....

....

....

....

   c   Adaptation

....

....

....

....

....

Exam-style set 1

1 Outline **one** cause of aridity in hot desert areas. (AO1)   4 marks

5

....

....

....

....

....

....

....

2 Using only information from the text below, analyse the various adaptations made by desert plants in order to survive the harsh conditions. (AO3)   6 marks

7

Plants in hot deserts are mainly low woody shrubs and trees. Leaves are 'replete' (fully supported with nutrients) and water-conserving. They tend to be small, thick and covered with a thick cuticle (outer layer). In cacti, the leaves are reduced to spines and photosynthesis is restricted to the stems. Cacti also depend on chlorophyll in the outer tissue of their skin and stems to conduct photosynthesis for the manufacture of food. Spines protect the plant from animals, shade it from the sun and collect moisture. Extensive shallow root systems are usually radial and, because they store water in the core of both stems and roots, they survive long periods of drought.

Some plants open their stomata only at night. These plants include yuccas, ocotillo and prickly pears.

The large numbers of spines on plants in desert areas shade the plant's surface. The same is true of the hairs on the woolly desert plants and many plants have silvery or glossy leaves.

Phreatophytes, like the mesquite tree, have adapted to desert conditions by developing extremely long root systems. The mesquite's roots are considered the longest of any desert plant and have been recorded as long as 20 metres.

Other plants survive by becoming dormant during dry periods, then springing to life when water becomes available. After rain falls, the ocotillo, for example, quickly grows a new suit of leaves, flowers bloom within a few weeks, and when seeds become ripe and fall, the ocotillo loses its leaves again and re-enters dormancy. This process may occur as often as five times a year.

....

....

....

....

....

....

....

....

....

....

....

3   Using Figure 2.10 and your own knowledge, assess the extent of the influence of
    water on the development of the landscape shown. (AO1, AO2)   6 marks

Figure 2.10 The Grand Canyon, Arizona, USA

....

....

....

....

....

....

....

....

....

4 Assess the extent to which population change, both locally and globally, has contributed to desertification. (AO1, AO2) **20 marks**

Plan and write your answer on a separate sheet of paper and keep it with your workbook.

Exam-style set 2

1 Outline **two** processes by which wind erodes desert surfaces. (AO1) **4 marks**

....

....

....

....

....

....

....

....

....

2 Using Figure 2.11 (a) and (b) and your own knowledge, explain why the Atacama area of Peru and Chile is a hot desert. (AO1, AO2) **6 marks**

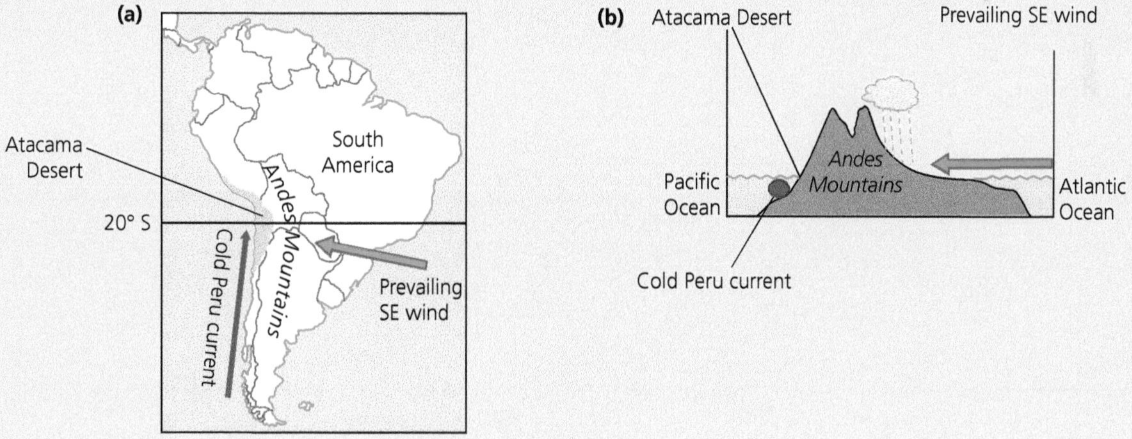

Figure 2.11 Reasons why the Atacama area is a hot desert

....

....

....

....

....

....

3 Analyse the climate graphs in Figure 2.12. (AO3) **6 marks**

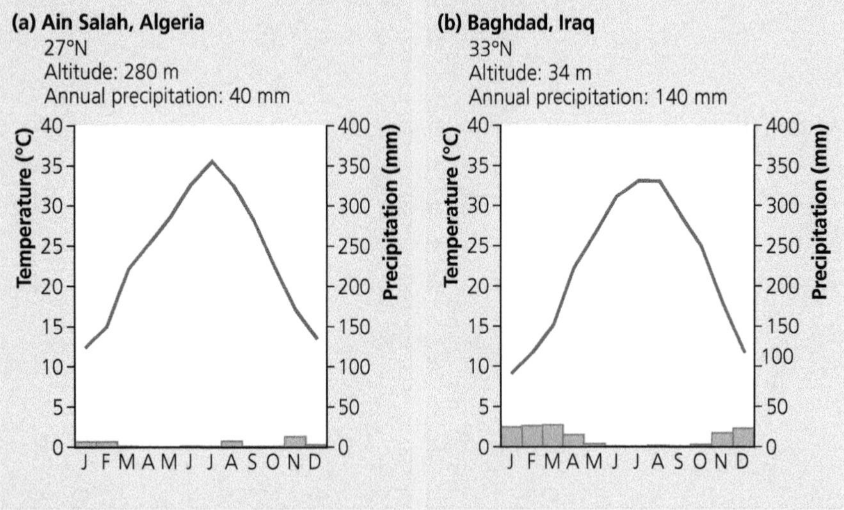

(a) Ain Salah, Algeria
27°N
Altitude: 280 m
Annual precipitation: 40 mm
Temperature (°C)
Precipitation (mm)
40 35 30 25 20 15 10 5 0
400 350 300 250 200 150 100 50 0
J F M A M J J A S O N D
(b) Baghdad, Iraq
33°N
Altitude: 34 m
Annual precipitation: 140 mm
Temperature (°C)
Precipitation (mm)
40 35 30 25 20 15 10 5 0
400 350 300 250 200 150 100 50 0
J F M A M J J A S O N D

Figure 2.12 Climate graphs for two hot desert areas

4 'The role of water is always the most important factor in the development of desert landforms.'

To what extent do you agree with this view? (AO1, AO2) **20 marks**

Plan and write your answer on a separate sheet of paper and keep it with your workbook.

Additional extended prose questions

Plan and write your answers on separate sheets of paper and keep them with your workbook.

1 For a local area where desertification has occurred, assess the impacts on people's lived experience of the place. (AO1, AO2) **20 marks**

2 'The aridity of hot desert areas is caused by a combination of several factors.'

To what extent do you agree with this statement? (AO1, AO2) **20 marks**

Topic 3 Coastal systems and landscapes

Coasts as natural systems

Like all systems in physical geography, coastal systems have inputs, stores, processes, flows and outputs.

When there is a balance between the inputs and outputs, the system is said to be in a state of dynamic equilibrium. If one of the elements of the system changes, it can upset this equilibrium. This is called feedback.

Positive feedback leads to further change in the coastline whereas negative feedback acts to stabilise coastal morphology and maintain a dynamic equilibrium. These mechanisms act together to form a rich variety of component landforms and landscapes.

A natural **landform** is a distinctive shape or feature found on the Earth's surface. A natural landscape is made up of a collection of landforms. Coastal landscapes are formed by a combination of erosional and depositional processes acting where sea and land meet.

Practice questions ?

1 Complete the following paragraph by filling in the gaps from the list below.

Inputs into a coastal system include:

- energy from _____, _____, _____ and _____.

- sediment either eroded from the local coastal rocks or transported from either _____ or along the _____

- the _____ and _____ of the local geology

- sea-level change

Erosional processes such as _____ and _____ result in erosional coastlines and landscapes as well as eroded material. _____ alters the eroded material by making it _____ and less _____.

Sediment is transported by either _____ or _____ until it reaches a low _____ environment where it is _____.

The outputs from the system are in the form of:

- _____ dissipated by breaking along the shoreline

- sediment accumulation above _____

- sediment moved on to other _____

List:

finer   hydraulic action   ocean currents   abrasion   water   energy   wind
wave energy   deposited   sediment cells   waves   offshore   attrition
high tide level   coast   rock type   wind   structure   angular   tides

2   Complete Figure 3.1 by choosing from the list below to show how negative feedback stabilises a coastal headland.

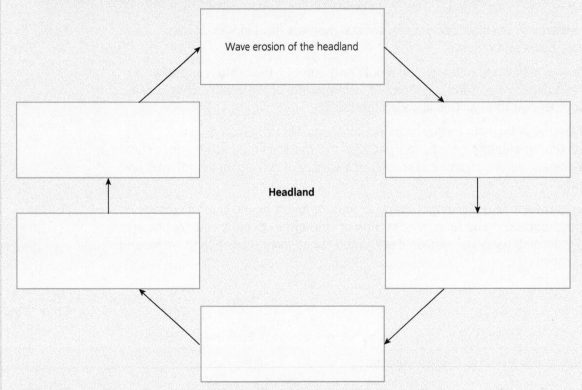

Figure 3.1 A system diagram showing negative feedback at a coastal headland

List:

Increasing loss of wave energy as the wave-cut platform widens

Cliff retreat

Undercutting and collapse of the cliffs

Rate of erosion reduced

Wave energy lost to friction as they cross the wave-cut platform

3   Design your own system diagram to illustrate the effect of positive feedback on a named stretch of coastline.

4   Describe a coastal landscape you have studied. Include in your description the landforms that make up that landscape.

....

....

....

....

....

....

....

Systems and processes

Energy transferred from wind to water creates waves and provides the energy to transport material on beaches and sand dunes. Waves have little impact on landscapes until they meet land and the sea shallows. Breaking waves are divided into two broad types, constructive and destructive.

A sea current is the permanent or seasonal movement of water. Ocean currents rarely have any effect on coastal landscapes. The main current to affect coastlines is the longshore current, giving rise to the transport process of longshore drift.

Tides are a periodic rise and fall in the depth of the sea caused by the gravitational interaction of the Earth, sun and moon.

Coastal sediment budget is the balance between what sediment goes into a stretch of coastline and what sediment comes out.

Weathering, the breakdown of rocks *in situ*, occurs on coastlines. There are three types of weathering: mechanical/physical, biological and chemical. Sea water is slightly alkaline and so is not generally able to react with carbonate rocks, though carbonation does occur to coastal limestones. Once weakened, the rocks are susceptible to erosion and mass movement.

Erosion occurs where waves hit coastlines. The force of the water itself or the load that a wave carries wears away coastlines. The debris created is transported by the sea water or, if small enough, by wind on a beach. Eventually deposition occurs when the water or air either loses energy or becomes less turbulent.

Practice questions ?

5   Outline how energy is transferred from wind to sea water.

....

....

....

....

....

6   Use labelled diagrams to show the difference between a constructive wave and a destructive wave.

7   Describe and explain the different impacts that constructive waves and destructive waves have on a beach.

....

....

....

....

....

....

....

....

8   Table 3.1 gives the percentage wind direction for Newquay, Cornwall.

Table 3.1

Direction	Percentage	Direction	Percentage
N	6.2	S	6.7
NNE	3.7	SSW	7.0
NE	2.9	SW	6.8
ENE	2.4	WSW	10.5
E	3.9	W	9.6
ESE	5.2	WNW	8.1
SE	5.2	NW	8.7
SSE	6.3	NNW	6.6

Complete Figure 3.2 to construct a wind rose diagram for Newquay.

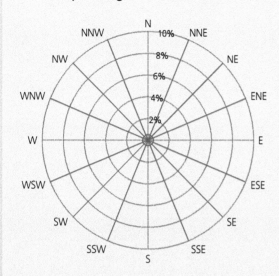

Figure 3.2 A wind rose diagram showing the mean annual percentage of wind direction for Newquay

9 Using Figures 3.2 and 3.3, state which of the named beaches around Newquay is likely to be the best for surfing. Give reasons for your choice.

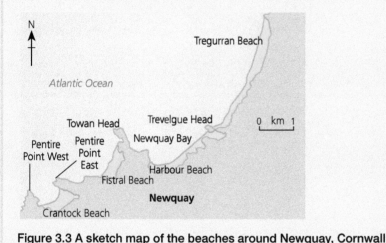

Figure 3.3 A sketch map of the beaches around Newquay, Cornwall

....

....

....

....

....

....

....

10 Define the term 'sediment cell'. Describe a sediment cell that you have studied.

....

....

....

....

....

....

....

....

11 Outline where coastal sediment comes from.

....

....

....

....

....

12 The following are all forms of coastal erosion. For each one, explain how it works and name one impact it has on coastlines.

a   Hydraulic action

....

....

....

....

....

b   Wave quarrying/cavitation

....

....

....

....

....

c   Corrasion/abrasion

....

....

....

....

....

....

d   Attrition

....

....

....

....

....

13 Wind moves lighter material on a beach by the processes of saltation and traction. Describe these processes.

....

....

....

....

....

....

14 Waves and sea currents move material by suspension as well as by saltation and traction. Define the term 'suspension'.

....

....

....

....

15 One of the commonest forms of marine sediment transport is longshore drift. Draw an annotated diagram to explain the process of longshore drift along a named stretch of coastline.

16 What causes freeze–thaw action along coastlines and why is it quite rare on British coastlines, especially along the south coast of England?

....

....

....

....

....

....

....

....

17 Apart from freeze–thaw action, name one other type of mechanical weathering process that occurs specifically on coastlines. Describe the process and explain how it affects the coastal landscape.

....

....

....

....

....

....

....

....

18 Using examples, explain the process of biological weathering on coastlines.

....

....

....

....

....

....

19 Under what circumstances does carbonation occur on coastlines?

....

....

....

....

....

20 What do you understand by the term 'mass movement'? Name **three** types of mass movement that occur on coastlines.

....

....

....

....

21 Using Figure 3.4 and an example you have studied, explain the process of rotational slumping for that location.

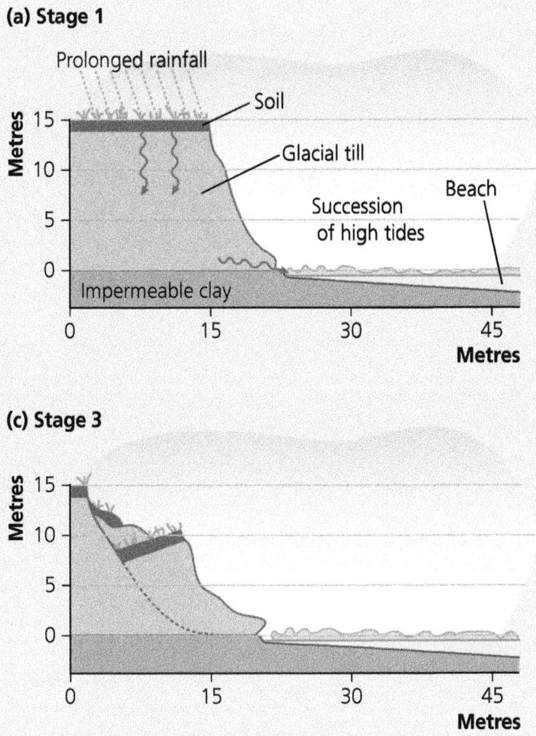

(a) Stage 1

(b) Stage 2

(c) Stage 3

Figure 3.4 Rotational slumping

....

....

....

....

....

Coastal landscape development

Coastal landforms and landscapes can be categorised into the following types:

- **Landforms and landscapes of coastal erosion.** These include cliffs and wave-cut platforms, and cliff profile features including caves, arches and stacks.

- **Landforms and landscapes of coastal deposition.** These include beaches, spits, tombolos, offshore bars, barrier beaches and islands, and sand dunes.

- **Estuarine mudflat/saltmarsh environments and associated landscapes.**

Sea levels are not static. Global sea levels have risen approximately 120 m since the height of the last ice advance 18,000 years ago. This has given rise to 'drowned' coastlines, including rias, fjords and Dalmatian coasts.

Some coastlines have risen locally, emerging from under the water. This forms emergent landforms, including raised beaches and marine platforms.

Predicted climate change (global warming and increasingly extreme weather events) could eventually lead to rising sea levels with greater coastal flooding and erosion, particularly along coastlines in the path of tropical storms.

Practice questions ?

22 Explain how the process of abrasion forms wave-cut notches at the base of cliffs.

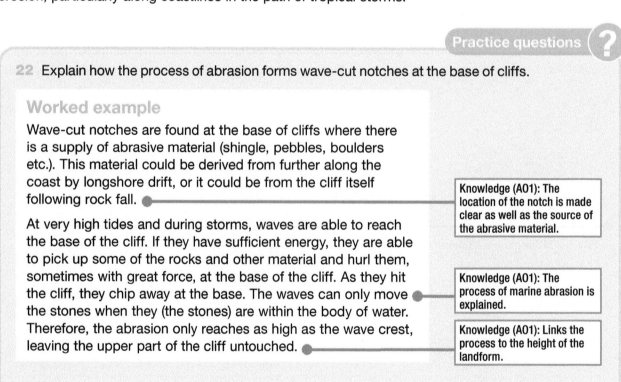

Worked example

Wave-cut notches are found at the base of cliffs where there is a supply of abrasive material (shingle, pebbles, boulders etc.). This material could be derived from further along the coast by longshore drift, or it could be from the cliff itself following rock fall.

> Knowledge (AO1): The location of the notch is made clear as well as the source of the abrasive material.

At very high tides and during storms, waves are able to reach the base of the cliff. If they have sufficient energy, they are able to pick up some of the rocks and other material and hurl them, sometimes with great force, at the base of the cliff. As they hit the cliff, they chip away at the base. The waves can only move the stones when they (the stones) are within the body of water. Therefore, the abrasion only reaches as high as the wave crest, leaving the upper part of the cliff untouched.

> Knowledge (AO1): The process of marine abrasion is explained.

> Knowledge (AO1): Links the process to the height of the landform.

23 Figure 3.5 shows caves, arches and a wave-cut platform on the Yorkshire coast.

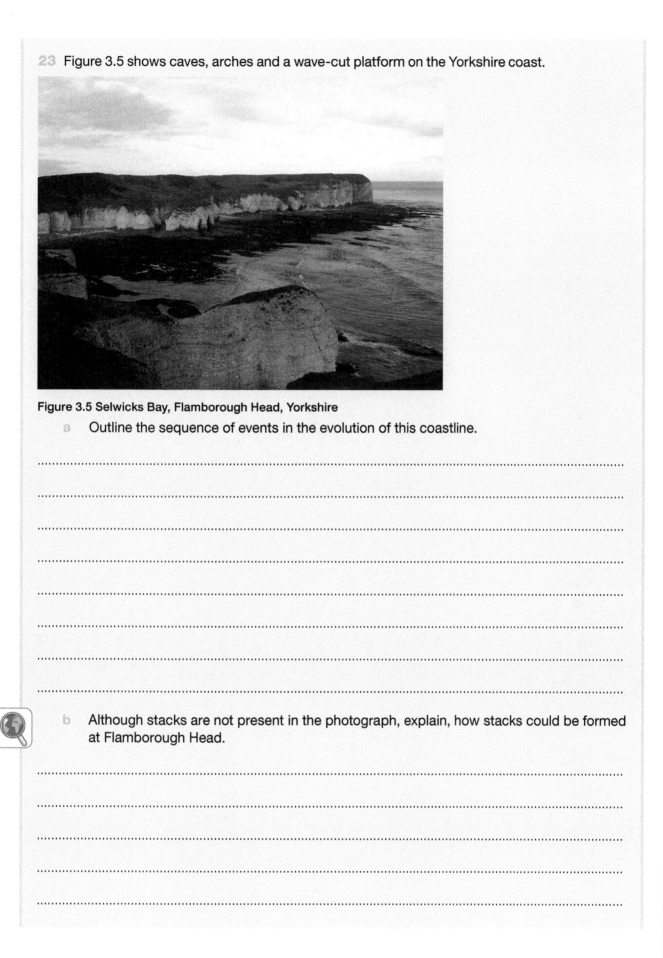

Figure 3.5 Selwicks Bay, Flamborough Head, Yorkshire

a   Outline the sequence of events in the evolution of this coastline.

....

....

....

....

....

....

....

....

b   Although stacks are not present in the photograph, explain, how stacks could be formed at Flamborough Head.

....

....

....

....

....

24 Annotate Figure 3.6 with the following labels:

Ripples   Berms   Beach cusps   Runnel   Ridge   Storm beach

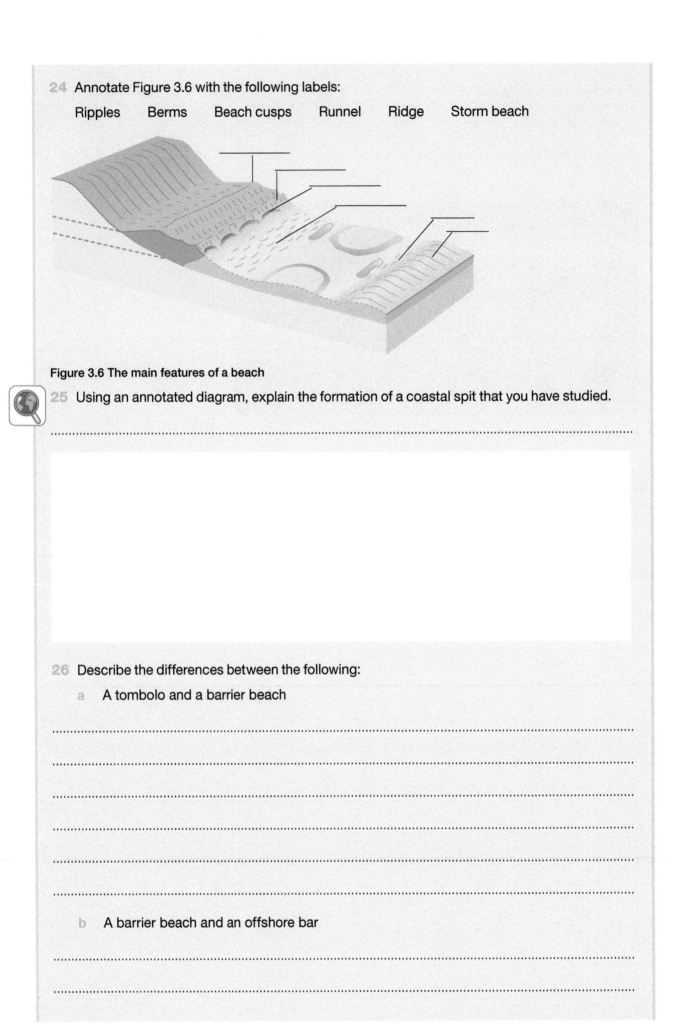

Figure 3.6 The main features of a beach

25 Using an annotated diagram, explain the formation of a coastal spit that you have studied.

....

26 Describe the differences between the following:

a   A tombolo and a barrier beach

....

....

....

....

....

....

b   A barrier beach and an offshore bar

....

....

27 Describe the main features of coastal sand dunes and explain the role of wind and vegetation in their formation.

Worked example

Coastal dunes are accumulations of sand that lie just inland from a beach. There are parallel ridges of sand of differing heights. The first ridge is often a temporary embryo dune that is easily destroyed by very high tides. It is less than 1 m high and is often totally bare of vegetation. Moving inland is a mobile foredune at around 5 m high. Here about 20% of the sand is exposed. The dune is yellow in colour and is alkaline. Further inland still are the grey, fixed dunes that grow up to 10 m high in the UK. They are 90% covered in vegetation and have a neutral pH. The oldest and furthest inland dunes become lower and more acidic, with 100% vegetation cover. Within the dunes there are sometimes 'dune slacks'. These are depressions that extend down below the water table and so are damp, sometimes occupied by small ponds.

> Knowledge (A01): The first paragraph is all description. It includes information on location, height, colour, acidity and vegetation cover, answering the first part of the question thoroughly.

Dunes are formed where there is a dominant onshore wind, blowing across wide, low-angled sandy beaches. The wind moves the dry, finer sand grains by traction (rolling) and saltation (bouncing) until they are trapped by a piece of debris or hardy plants right at the top of the beach. As more sand gathers, other plants colonise the dune. They slow down the wind speed, allowing further deposition of sand. Their deep, complex roots hold the sand together. As the dunes get older, plant matter accumulates in the sand, turning it grey. Sometimes some of the vegetation is destroyed and wind blows away large amounts of sand to form a blow-out.

> Application (A02): The role of wind is explained.

> Application (A02): The role of vegetation is explained.

> Knowledge (A01): An anomaly is identified, showing that the student understands more than just a textbook model.

28 Using an example you have studied, describe and explain the formation of a saltmarsh and tidal (mud) flats.

....

....

....

....

....

....

....

....

....

29 Complete the graph in Figure 3.7 from the data in Table 3.2. Describe and account for the changes in sea level over the last 18,000 years.

Table 3.2

Years before present	Approximate sea level compared to present day/m
18,000	−120
16,000	−115
14,000	−94
12,000	−75
10,000	−63
8,000	−20
6,000	−2
4,000	−1.5
2,000	+0.5
Present day	0

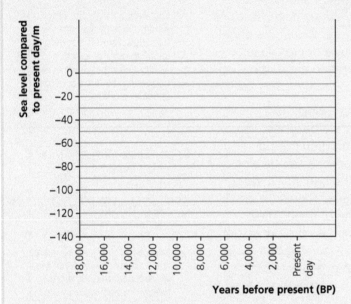

Figure 3.7 Changing sea levels over the last 18,000 years

....

....

....

....

....

....

....

....

30 Study Figure 3.8. Describe the coastline and assess the role played by sea-level change in the formation of this coastline.

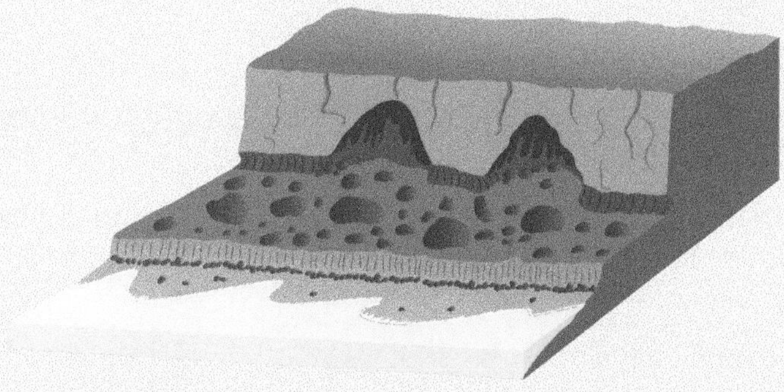

Figure 3.8 A sketch of part of the coastline of the Isle of Arran

....

....

....

....

....

....

31 Draw an annotated sketch of either a ria or a fjord or a Dalmatian coastline, describing the main features and explaining the formation of the main features.

Coastal management

Securing the sustainable use of the coastal zone presents particular challenges to landowners and/or coastal managers. These include the often complex issues related to whose responsibility it is to manage the coastline. An estimated 60% of the world's human population live on or close to the coast and the pressures on coastal environments for economic development are particularly high.

A range of protection and management strategies have been put in place to:

- provide defence and protection from coastal flooding
- provide protection from coastal erosion and its impacts

Other considerations for managers may be:

- the stabilisation of beaches or dunes as an amenity and tourism asset
- the protection of fragile yet important ecosystems

The types of strategies employed include the following:

- Hard engineering — defined as controlled disruption of natural processes by using human-made structures. These structures include sea walls, groynes and breakwaters.
- Soft engineering — the use of natural systems for coastal defence, such as beaches and dunes which can absorb and adjust to wave and tide energy. It involves manipulating and maintaining these systems, without changing their fundamental structures.

Sustainable approaches to coastal management include:

- shoreline management plans (UK)
- integrated coastal zone management (United Nations)

32 Choose **two** types of hard engineering strategy. For each, explain their specific purpose and describe how they work.

Strategy 1 ....

....

....

....

Strategy 2 ....

....

....

....

33 Using an example of a scheme of soft engineering you have studied, describe the scheme and assess the extent to which it is sustainable.

....

....

....

....

....

....

....

....

34 Describe the coastal management issue shown in Figure 3.9. Suggest one way in which this could be managed and assess the sustainability of the scheme.

Figure 3.9 The promenade, Crosby Beach

....

....

....

....

....

....

....

....

35 For a stretch of coastline you have studied, analyse the extent to which the defences have achieved the goals set by coastal managers.

....

....

....

....

....

....

....

....

Exam-style set 1

5

1   Outline **two** weathering processes that take place on coastlines. (AO1)   4 marks

....

....

....

....

....

....

....

7

2   Study Figure 3.10 (a) and (b). Using the figure and your own knowledge, assess
     the extent to which different types of waves have different effects on beaches.
     (AO1, AO2)   6 marks

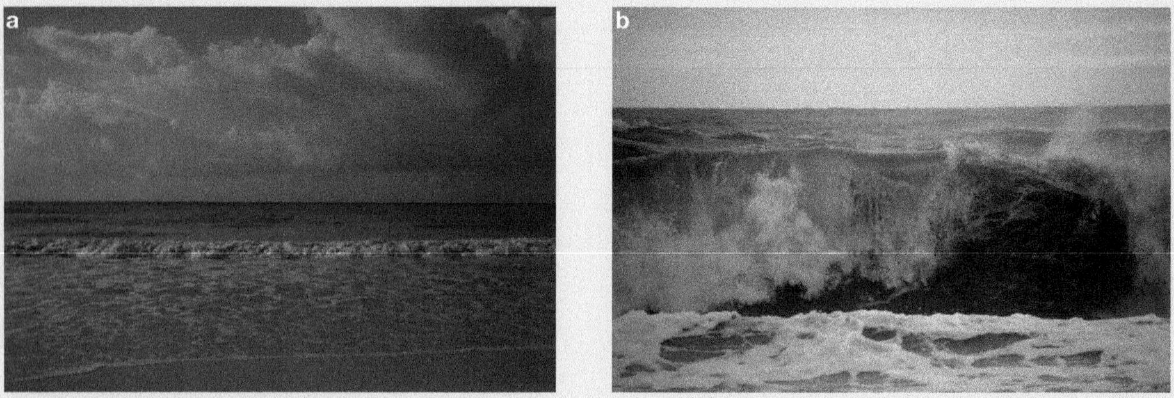

Figure 3.10 Contrasting wave types

....

....

....

....

....

....

....

....

....

....

3   Study Figures 3.11 and 3.12. Analyse the information shown. (AO3)   6 marks ⑦

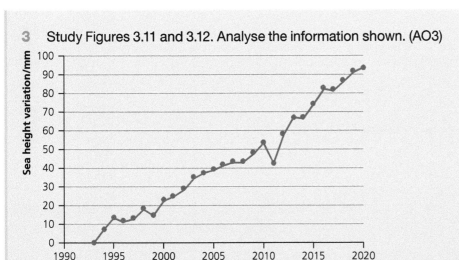

Figure 3.11 The variation in sea height from a base level set in 1993

Data from https://sealevel.nasa.gov/understanding-sea-level/key-indicators/global-mean-sea-level

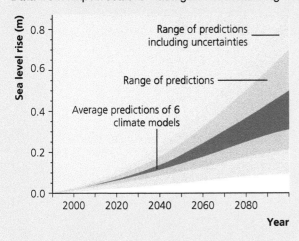

Figure 3.12 Range of possible predicted sea-level rise

Source: NASA Earth Observatory

....

....

....

....

....

....

....

....

....

4   'Climate change presents both risks and opportunities for human occupation of coastlines.'

With reference to a named coastal landscape beyond the UK, to what extent do
you agree with the above statement? (AO1, AO2)   20 marks ㉕

Plan and write your answer on a separate sheet of paper and keep it with your workbook.

1   Explain the concept of eustatic sea-level change. (AO1)   4 marks

....

....

....

....

....

....

....

2   Some students carried out fieldwork on the shingle beach and ridge at Porlock in North
    Devon, shown on the sketch map in Figure 3.13. One of the characteristics that they
    measured was the roundness of the pebbles at the western end (Gore Point) and the
    eastern end (Hurlstone Point) of the beach.

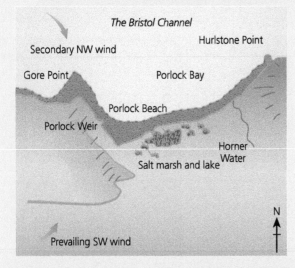

Figure 3.13 A sketch map of Porlock Bay and Beach

    A sample of 150 pebbles was collected at each point and the Cailleux roundness index was
    measured, where 0 is completely angular and 1,000 is a perfect sphere. The results are
    shown in Table 3.3.

Table 3.3

Shape	Gore Point	Rank	Hurlstone Point
0–100	2	10	1
101–200	14	5	2
201–300	28	2	8
301–400	36	1	25
401–500	25	3	23
501–600	19	4	16
601–700	9	6	26
701–800	4	9	27
801–900	8	7	15
901–1,000	5	8	7

On the axes provided (Figure 3.14), construct two histograms, one for Gore Point and one for Hurlstone Point, to show the frequency of different pebble roundness measurements at each end of Porlock Beach. Analyse your completed graphs. (AO3)   6 marks

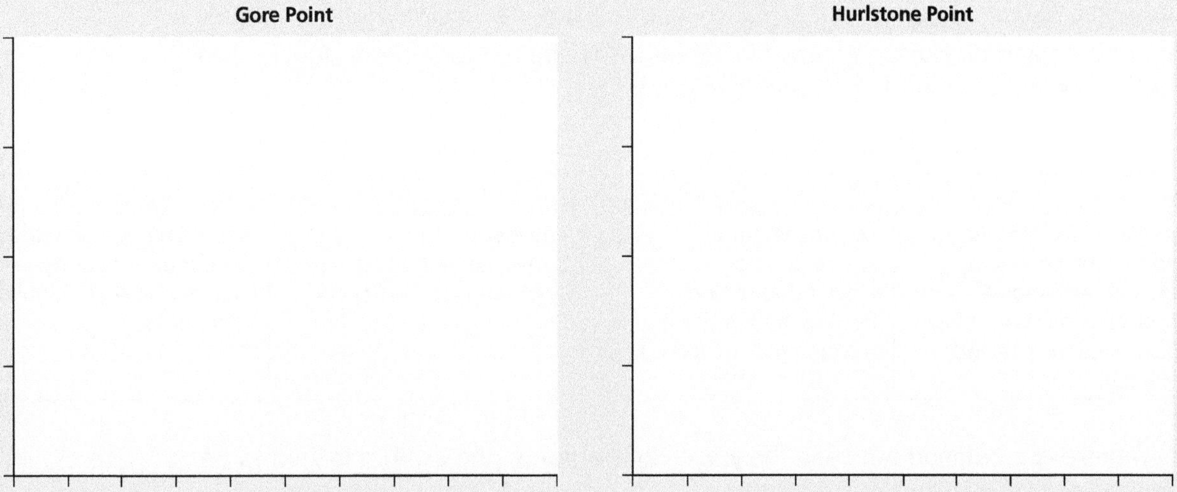

Gore Point

Hurlstone Point

Figure 3.14

....

....

....

....

3   Using Figure 3.5 (page 49) and your own knowledge, assess the role of erosion in the formation of the coastal landscape shown. (AO1, AO2)   6 marks

....

....

....

....

....

....

4   To what extent does increasing globalisation affect the sustainable management of coasts? (AO1, AO2)   20 marks

Plan and write your answer on a separate sheet of paper and keep it with your workbook.

Additional extended prose questions

Plan and write your answers on separate sheets of paper and keep them with your workbook.

1   'Sustainable solutions to coastal management require a variety of approaches.'

To what extent do you agree with this statement? (AO1, AO2)   20 marks

2   Assess the importance of the role played by changing sea levels in the formation of differing coastal landscape types. (AO1, AO2)   20 marks

Topic 4 Glacial systems and landscapes

Glaciers as natural systems

Like all systems in physical geography, glacial systems have inputs, processes, components and outputs, as shown in Table 4.1. These act together to form a rich variety of component landforms and landscapes.

Table 4.1 Parts of the glacial system

Inputs	Processes	Components	Outputs
Energy (solar radiation and geothermal heat) Mass (direct snowfall and other precipitation, blown snow, avalanches and rock fall)	Ice movement Glacial erosion Glacial deposition Melting, sublimation and calving erosion processes	The ice Erosional landforms Depositional landforms	Water (meltwater, water vapour and icebergs) Glacial and fluvioglacial sediments

When there is a balance between the inputs and outputs, the system is said to be in a state of dynamic equilibrium. If one of the elements of the system changes, it can upset this equilibrium. This is called feedback. These mechanisms act together to form a rich variety of component landforms and landscapes.

A natural **landform** is a distinctive shape or feature found on the Earth's surface. A natural landscape is made up of a collection of landforms. Glacial, fluvioglacial and periglacial landscapes are formed by a combination of erosional and depositional processes acting within cold environments.

Practice questions ?

1 Complete Figure 4.1 by choosing from the list below to show how positive feedback can accelerate the melting of Arctic ice.

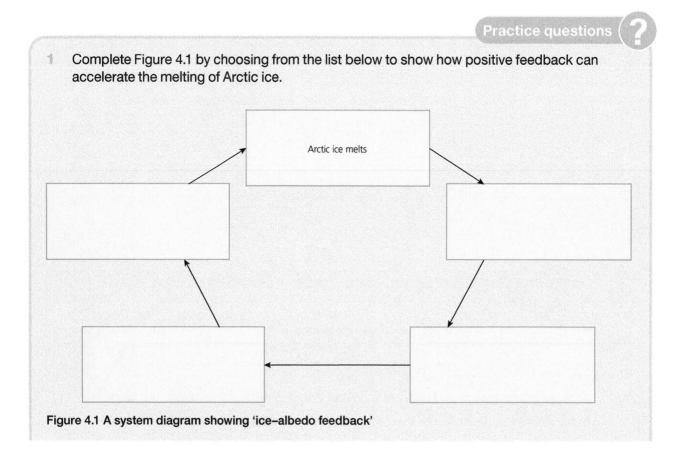

Figure 4.1 A system diagram showing 'ice–albedo feedback'

List:

These darker and duller surfaces have a lower albedo

More solar radiation is absorbed, causing the surface to warm

Land and water surfaces are exposed

Exposed land and sea reflect less solar radiation than ice

2   Design your own system diagram to illustrate the effect of negative feedback in a cold environment.

3   Describe a periglacial landscape you have studied. Include in your description the landforms that make up that landscape.

....

....

....

....

....

....

....

....

4   Outline the main differences between a periglacial landscape and an Alpine glacial environment.

....

....

....

....

....

The nature and distribution of cold environments

Both present and past cold environments may be described as polar, alpine, glacial and periglacial. At the height of the last ice advance, glacial areas were much more extensive than they are today. Polar conditions extended much further towards the equator than today. When the ice retreated, relict landscapes were left behind. In the UK, the north is dominated by areas of both upland and lowland which were shaped by glaciers and ice sheets and then modified by post-glacial processes.

The ice did not advance as far as the south of England. Instead, it experienced a tundra climate with associated periglacial conditions. Tundra environments have long winters and cool summers, allowing only limited vegetation growth. This, coupled with the high acidity and impermeability of the subsurface permafrost, leads to a distinctive soil type with extremely low productivity. Tundra vegetation is composed of dwarf shrubs, sedges and grasses, mosses and lichens, with small areas of scattered trees.

There are no soils or vegetation in glacial environments: the ice scrapes any loose material away as it moves over the surface.

Practice questions ?

5   Using Figure 4.2 and your own knowledge, describe the following cold environments and their global distribution:

a   Glacial areas

....

....

....

....

....

....

b   Alpine areas

....

....

....

....

....

....

c   Tundra areas

....

....

....

....

....

....

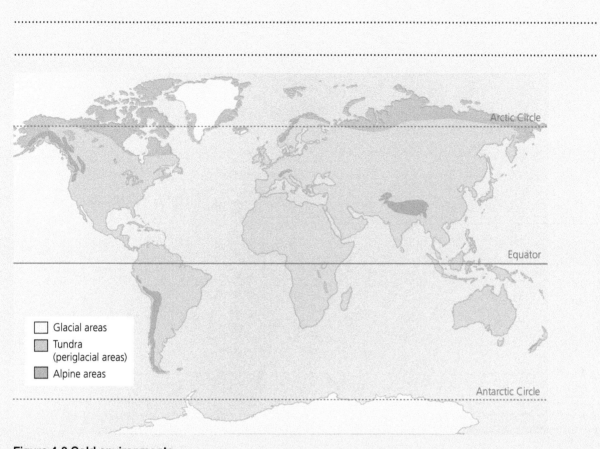

Figure 4.2 Cold environments

6 Outline the differences between the distributions shown in Figure 4.3 and those of the present day.

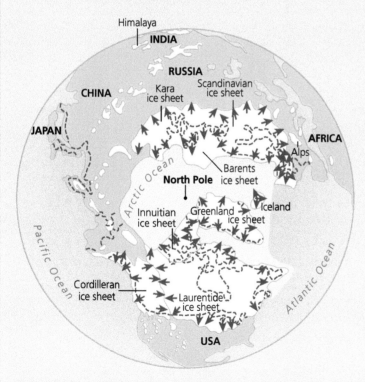

Figure 4.3 The distribution of continental ice sheets in the northern hemisphere at the height of the Pleistocene epoch

....

....

....

....

....

....

....

....

7 Describe the climates for Eismitte, Greenland (Figure 4.4a), and for Barrow, Alaska, USA (Figure 4.4b). Outline the differences between them.

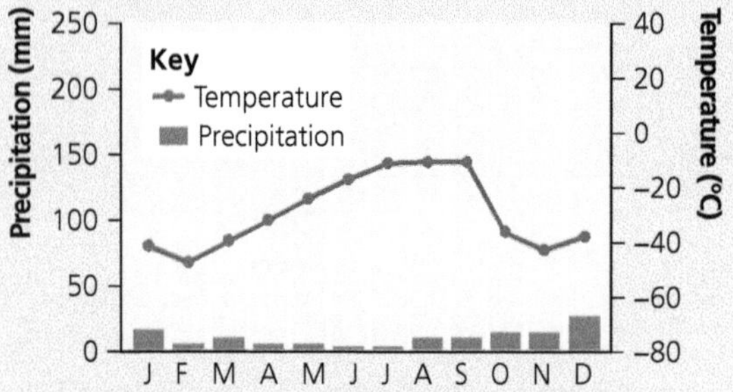

Figure 4.4a Mean monthly temperature and precipitation values for Eismitte, Greenland, a polar location

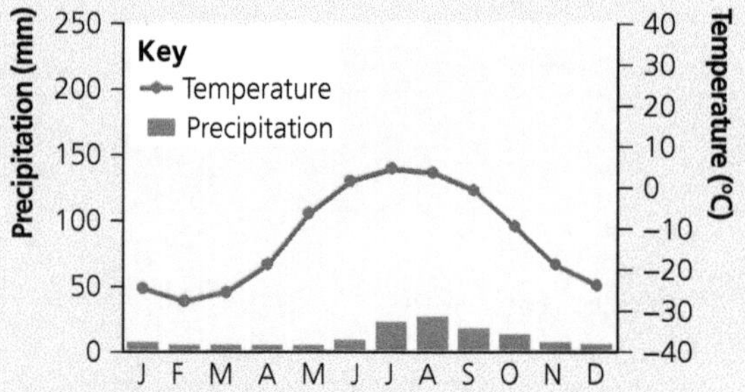

Figure 4.4b Mean monthly temperature and precipitation values for Barrow, Alaska, USA, a tundra location

....

....

....

....

....

....

8 Complete the soil profile of a tundra soil (Figure 4.5) by linking the labels provided to the correct part of the profile.

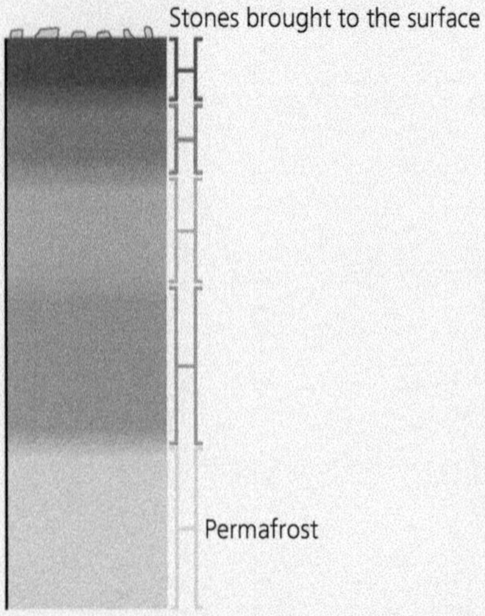

Stones brought to the surface

Permafrost

- Lack of clearly differentiated soil horizons (layers), caused by lack of soil biota to mix layers
- A thin surface organic layer which is often very acidic
- Waterlogged in summer, as water unable to percolate into permafrost
- Gleyed (because of the waterlogging, iron compounds are reduced to their ferrous form, which is blue/grey)
- Impermeable frozen ground

Figure 4.5 The profile of a tundra soil

Systems and processes

A glacier is an open system with inputs, stores, transfers and outputs.

Glacial budgets are the balance between the inputs and outputs of a glacier. This can refer to annual or historical budgets.

If accumulation exceeds ablation, the net mass of the glacier increases. When ablation exceeds accumulation, the net mass decreases.

Glacial advance occurs when the ice mass has extended further downhill and covers more area. Glacial retreat indicates that the mass of ice is reduced and the glacier's snout is found at a greater (colder) altitude.

Warm-based glaciers are found in temperate and/or high-altitude regions. They are characterised by:

- high rates of accumulation in winter

- high rates of ablation in spring and summer, with large amounts of meltwater

- both basal and internal flow

- high rates of erosion and deposition

Cold-based glaciers are characterised by:

- low rates of accumulation

- low rates of melting

- slow, mainly internal movement

Glacial processes

Glacial landforms and landscapes are the result of a combination of several processes (see Table 4.2).

Table 4.2 Glacial processes

Weathering	Ice movement	Erosion	Transport	Deposition
Frost action Nivation Chemical weathering	Internal deformation Rotational flow Compressional and extensional flow Basal sliding	Plucking Abrasion	Subglacial Englacial Supraglacial	At the snout, the margins and beneath a glacier

Fluvioglacial processes

In summer, the surface ice of warm-based glaciers melts. The resulting meltwater flows on the surface, at the sides, inside and at the base of the glacier. It eventually emerges at the snout where the water slows down and deposition occurs.

Periglaciation and permafrost

Regions close to glaciated areas are dominated by extremely low temperatures for most of the year, with a short, slightly warmer summer. The ground below the top 0.5–3.0 metres is permanently frozen. The surface layer melts seasonally, becoming waterlogged. This allows the soil or weathered material to be churned up by the variation in temperatures and flow down gentle slopes.

9   Complete Figure 4.6, using the following labels:

- Area where inputs exceed outputs, leading to a net gain in mass
- The glacier itself, consisting of ice, water, air and rock debris
- Snow and ice from both precipitation and avalanche; rock debris from the valley side and floor
- Water vapour, calving, melting (to form meltwater streams) and rock debris
- The separation line between areas of net loss and net gain
- Area where outputs exceed inputs, leading to a net loss of mass
- Glacial movement

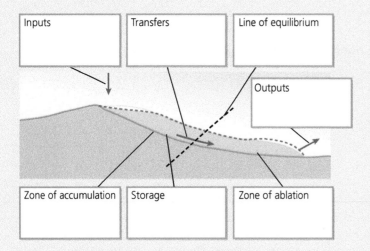

Figure 4.6 A glacial system or budget diagram

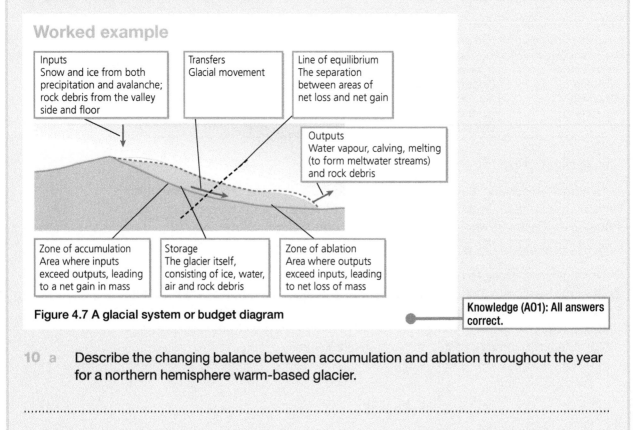

10 a   Describe the changing balance between accumulation and ablation throughout the year for a northern hemisphere warm-based glacier.

....

....

....

....

....

....

b Explain why the balance changes throughout the year.

....

....

....

....

11 a Describe the changes in the mass of mountain glaciers around the world, as shown in Figure 4.8.

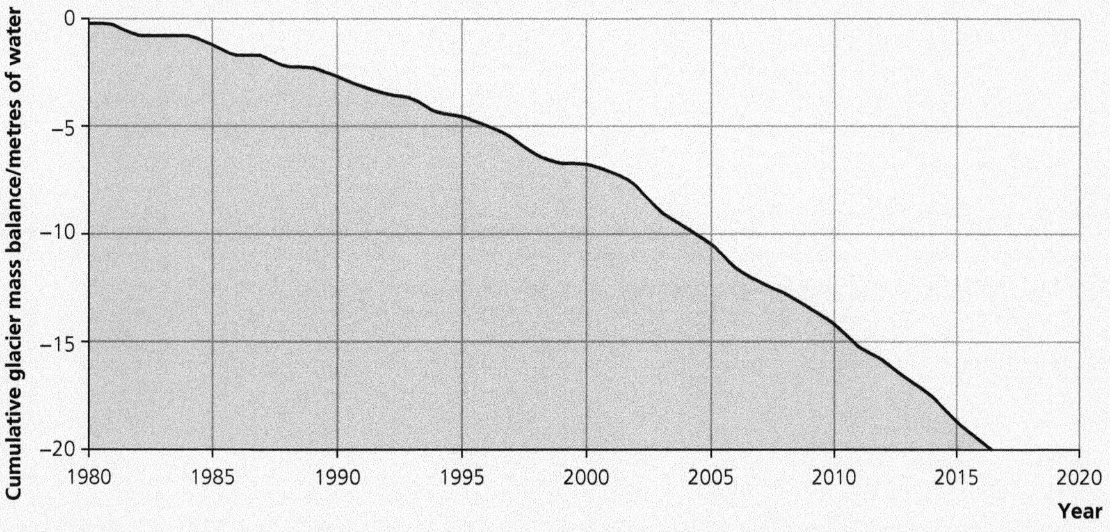

Figure 4.8 The cumulative mass loss in 'metres of water equivalent' (the depth of the meltwater spread out over the glacier's surface area) of global mountain glaciers

Source: NOAA Climate.gov, https://www.climate.gov/news-features/understanding-climate/climate-change-glacier-mass-balance

....

....

....

....

b Outline the effect this will have on global sea levels.

....

....

....

c Describe **two** problems this may cause for people living downstream of glaciated mountains.

....

....

....

....

....

12 Explain how long-term temperature changes have caused glacial advance and retreat throughout the Pleistocene epoch.

....

....

....

....

....

....

13 What are the main differences between warm- and cold-based glaciers?

....

....

....

....

....

....

....

....

....

14 In a glacial context, what are the differences between weathering and erosion?

....

....

....

....

....

....

15 Explain how frost action operates and how it influences the development of glacial landforms.

....

....

....

....

....

....

....

16 Explain how nivation operates and how it influences the development of corries.

....

....

....

....

....

....

....

....

....

17 Explain how the following types of ice movement operate.
   a   Internal deformation

....

....

....

....

....

b   Basal sliding and regelation flow

....

....

....

....

....

....

18 Using Figure 4.9, explain how extensional and compressional flow are related to changes in valley gradient.

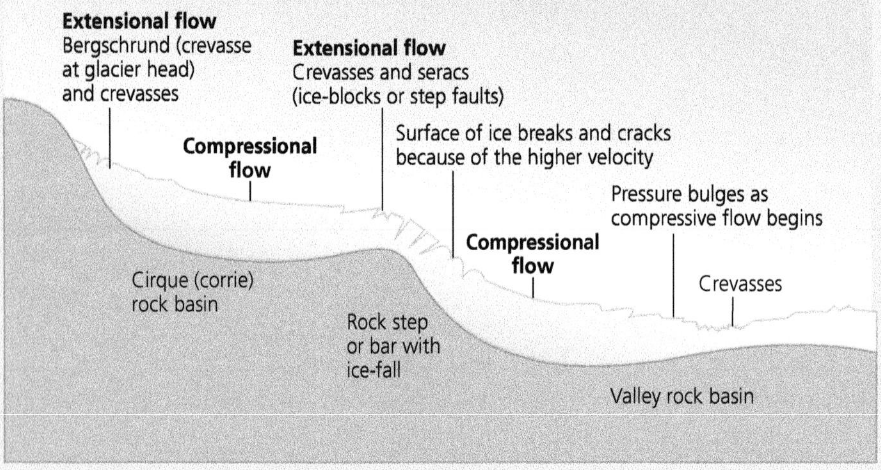

Extensional flow
Bergschrund (crevasse at glacier head) and crevasses

Extensional flow
Crevasses and seracs (ice-blocks or step faults)

Compressional flow

Surface of ice breaks and cracks because of the higher velocity

Pressure bulges as compressive flow begins

Compressional flow

Crevasses

Cirque (corrie) rock basin

Rock step or bar with ice-fall

Valley rock basin

Figure 4.9 Extensional and compressional flow

....

....

....

....

....

....

19 Rotational flow occurs in corries. It is a combination of both extensional and compressional flow. Draw a labelled diagram to show how ice moves in a corrie.

20 The two main processes of glacial erosion are plucking and abrasion. Explain how each of these processes works and how each influences the development of erosional landforms.

....

....

....

....

....

....

....

....

21 Explain why the load carried by meltwater streams is much more rounded than that carried by a glacier.

....

....

....

22 Explain the circumstances under which a glacial meltwater stream might deposit its load.

....

....

....

....

....

....

....

....

23 Using Figure 4.10, describe the distribution of permafrost in the northern hemisphere.

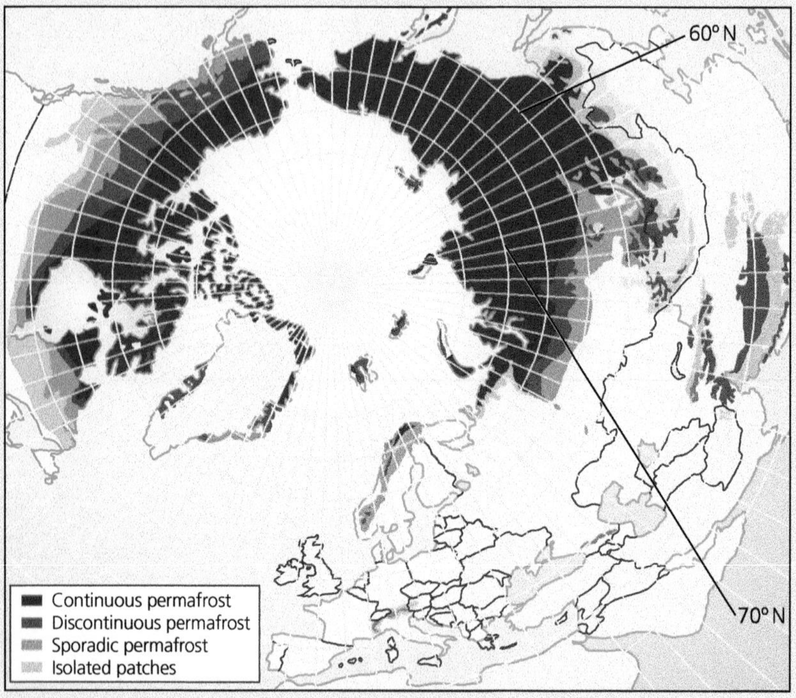

Figure 4.10 The distribution of permafrost in the northern hemisphere

Source: International Permafrost Association

....

....

....

....

....

....

....

....

24 Describe the variations in depth of the permafrost as shown in Figure 4.11.

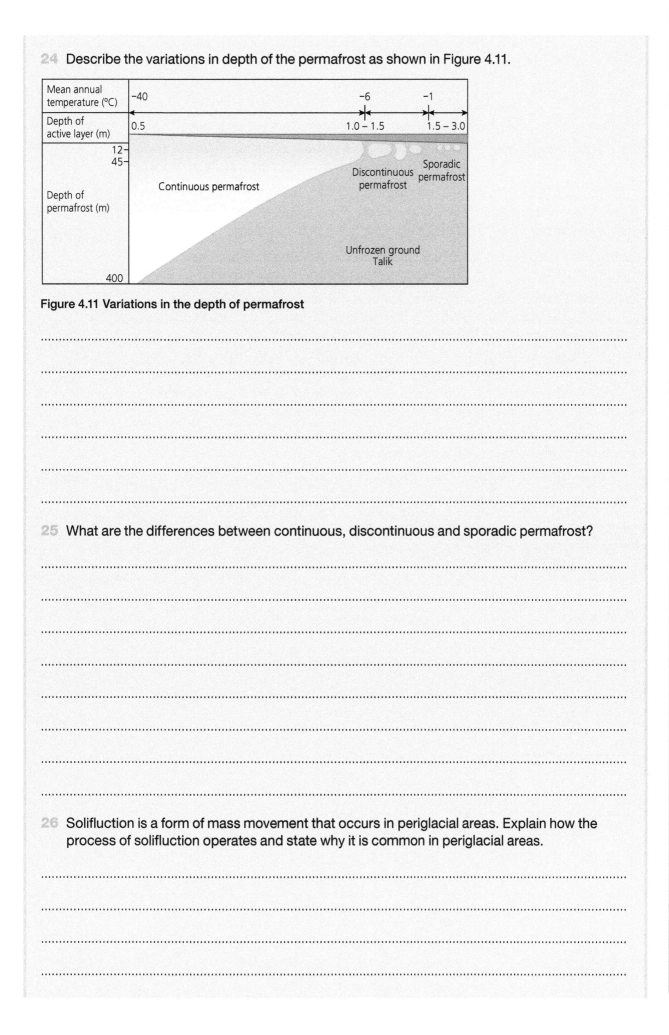

Figure 4.11 Variations in the depth of permafrost

....

....

....

....

....

....

25 What are the differences between continuous, discontinuous and sporadic permafrost?

....

....

....

....

....

....

....

....

26 Solifluction is a form of mass movement that occurs in periglacial areas. Explain how the process of solifluction operates and state why it is common in periglacial areas.

....

....

....

....

....

....

....

....

Glaciated landscape development

Glaciated landscapes consist of landforms that have developed from glacial erosion and deposition. They can be polar, mountainous, or temperate lowlands.

Fluvioglacial landforms and landscapes are those which have been formed by high-energy glacial meltwater. Where the stream energy is lost, deposition occurs, leading to features such as kames, eskers and outwash plains.

Periglacial landforms develop in areas where there are very cold winters and short, cool summers. On low-angled, bare rock surfaces, frost action occurs, leading to blockfields. Where there is a slope, scree is formed. Active layers thaw in summer and are able to flow down gentle slopes, forming terraces and lobes.

Patterned ground is an umbrella term that covers two completely different landforms that have some similarities in appearance:

- Stone polygon — formed by the process of frost heave (this type includes circles, nets and stripes).

- Ice wedge polygon — formed by the process of ground contraction.

Thermokarst is a term used to describe a landscape where there has been sporadic melting of the permafrost.

Practice questions ?

27 Using one or more diagrams, explain the formation of a corrie and associated arêtes and hanging valleys.

....

....

....

28 Describe a glacial trough and its associated features (hanging valleys and truncated spurs) that you have studied. Explain how they were formed.

....

....

....

....

....

....

....

....

....

....

29 Explain how a roche moutonnée is formed. Consider in your answer the roles of erosion processes and regelation flow.

....

....

....

....

....

....

....

....

....

....

30 Describe the landscape of a glaciated area located beyond the UK where erosion has been dominant.

Worked example

The Mont Blanc massif is an area of Alpine glaciation. It is an area of approximately 40 km by 16 km and includes the summit of Mont Blanc itself: one of the highest peaks in Europe at 4,807 m. The summit of Mont Blanc is permanently capped by snow and ice. Corrie and valley glaciers cover approximately 100 km^2 of the area. In the north of the massif there are more than 25 glaciers extending to lower altitudes, whereas on the southern flanks there are 25 glaciers near the summit but only four reach lower levels. The best-known glacier is the Mer de Glace which, at 7.5 km long and 200 m in depth, is the largest glacier in France. ●———————————

Knowledge (A01): Starts off well. It is clear that the account is about the Mont Blanc massif and not some generic glaciated area.

There are many corrie glaciers occupying hollows in the mountainside. Some former corrie glaciers have melted and left small armchair-shaped corries behind.

The glaciers occupy deep troughs, with high, steep, rocky sides and all the features of glacial erosion, including hanging valleys, truncated spurs (e.g. in the valley of the River Arve near Chamonix) and roches moutonnées (such as those below the snout of the Argentière Glacier). ●———————————

Knowledge (A01): Glaciated features located.

The area is increasingly being affected by climate change. The glaciers are retreating, exposing the flat valley floors. Because of this, erosion is not the only process. There are moraines on the valley floors and sides, as well as some meltwater features such as kame terraces. ●———————————

Application (A02): Shows understanding by bringing in other processes that also contribute to this landscape.

31 Study Figure 4.12. Describe the positions of the various types of moraine and explain how they were formed.

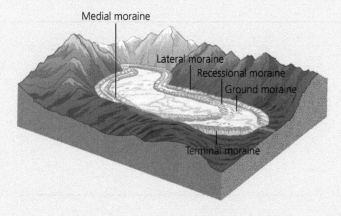

Figure 4.12 The relative positions of glacial moraines

....

....

....

....

....

....

....

....

32 Draw a labelled diagram of a drumlin.

33 Describe the various locations where meltwater channels can be found.

....

....

....

34 Using an example, explain how pro-glacial lakes are formed.

....

....

....

....

....

....

35 Study Figure 4.13. Explain how periglacial processes led to this blockfield (felsenmeer) being formed.

Figure 4.13 A blockfield (felsenmeer) on the Stiperstones, Shropshire

....

....

....

....

....

36 Outline the process of frost heave and show how it can lead to the formation of stone polygons.

....

....

....

....

....

....

....

....

....

....

37 Ice wedges and associated polygons only occur where the winter temperatures are extremely low, leading to ground contraction and the appearance of cracks in the surface. Explain how, over time, this leads to ice wedge polygons.

....

....

....

....

....

....

....

....

....

38 Choose **one** type of pingo (either an open- or a closed-system pingo) and draw a labelled sketch to describe its appearance. Explain how it has been formed.

....

....

....

....

....

....

....

....

....

Human impacts on cold environments

Despite the harshness of glacial and periglacial environments, they are very fragile. They are susceptible to changes brought about by human activity. These activities include:

- human-induced climate change
- pollution
- infrastructure development
- economic activity (tourism, mineral and oil/gas exploitation)
- commercial fishing and hunting

Traditional (hunter-gatherer) economies had little impact. They were small in scale and the landscape is extensive. Modern developments are generally on a much larger scale and much more damaging.

Recent climate change is causing damage through:

- a reduction in the size of ice caps and the loss of sea ice
- the melting of permafrost
- the possible future extinction of vulnerable species, e.g. the polar bear

It is also bringing some benefits through:

- the exposure of previously inaccessible mineral deposits
- the introduction of limited agriculture

Practice questions ?

39 Outline what is meant by 'environmental fragility'.

....

....

40 Unplanned development in permafrost areas can lead to the formation of thermokarst. Describe the nature of these developments and assess the extent to which they could lead to global climate change.

....

....

....

....

....

....

....

....

41 Extensive melting of permafrost can lead to the release of methane from the ground. Explain why this is harmful to the global environment.

....

....

....

....

Exam-style questions ❓

Exam-style set 1

1 Outline the ways in which a glacier transports material. (AO1) **4 marks** ⏱ 5

....

....

....

....

....

2 Using Figure 4.14 and your own knowledge, assess the view that human activity is not solely to blame for all glacial retreat. (AO1, AO2) **6 marks** ⏱ 7

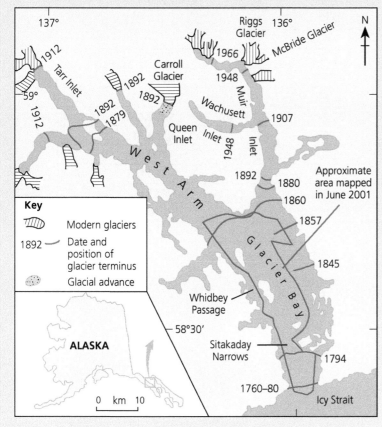

Figure 4.14 The locations of glacial termini in Glacier Bay, Alaska

3 Complete Table 4.3 and calculate the standard deviation of the corrie orientation using the formula:

$$SD = \sqrt{\frac{\sum(x - \bar{x})^2}{n}}$$

Analyse the completed data. (AO3)

6 marks

Table 4.3 The orientation (in degrees) of corries surrounding Ben Macdui, Cairngorm

Corrie orientation x	$x - \bar{x}$	$(x - \bar{x})^2$
70	−2.67	7.13
135	62.33	3,885.03
125	52.33	2,738.43
45	−27.67	765.63
0	−72.67	5,280.93
70	−2.67	7.13
90		
80	7.33	53.73
30	−42.67	1,820.73
70	−2.67	7.13
175	102.33	1,047.43
130	57.33	3,286.72
60	−12.67	160.53
0	−72.67	5,280.93
10	−62.67	3,927.53
$\sum x = 1,090$	$\bar{x} =$	$\sum(x - \bar{x})^2 =$
		Standard deviation =

4   Assess the extent to which fragile cold environments have been affected by globalisation. (AO1, AO2)

20 marks

Plan and write your answer on a separate sheet of paper and keep it with your workbook.

Exam-style set 2

1   Explain the concept of dynamic equilibrium when applied to glacial systems. (AO1)

4 marks

....

....

....

....

2   Using Figures 4.15, 4.16 and 4.17, analyse the trends shown. (AO3)

6 marks

Information relating to the melting of the Greenland Ice Cap

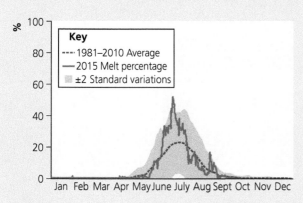

Figure 4.15

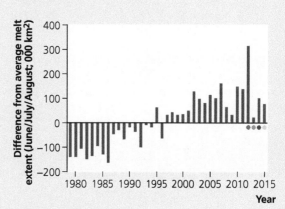

Figure 4.16

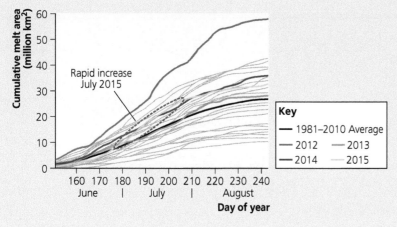

Figure 4.17

....

....

....

....

....

....

3   Using Figure 4.18 and your own knowledge, assess the relative importance of
    pre-glacial, glacial and post-glacial processes in creating the landscape shown
    in Figure 4.18. (AO1, AO2)   6 marks

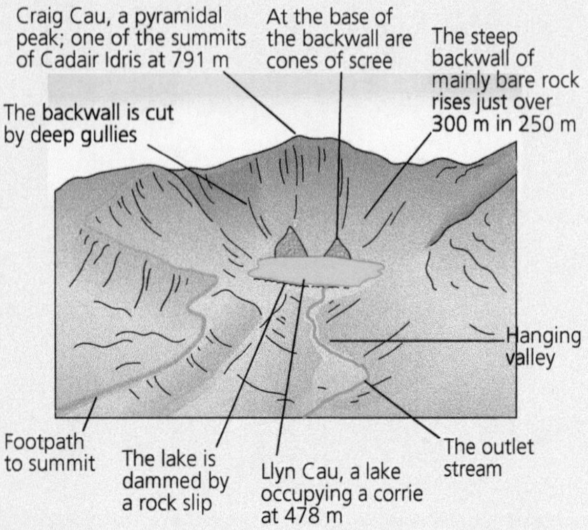

Figure 4.18 A field sketch of Llyn Cau, Cadair Idris

....

....

....

....

....

....

....

4   Using a case study of a contrasting glaciated landscape from beyond the UK,
    assess the extent to which climate change presents both challenges and
    opportunities for human occupation. (AO1, AO2)   20 marks

    Plan and write your answer on a separate sheet of paper and keep it with your workbook.

Additional extended prose questions

Plan and write your answers on separate sheets of paper and keep them with your workbook.

1   Assess the extent to which different processes have influenced the development of the
    landscape of a named glaciated environment at a local scale. (AO1, AO2)   20 marks

2   Evaluate the importance of meltwater in the development of fluvioglacial
    landscapes. (AO1, AO2)   20 marks

Topic 5 Hazards

The concept of hazard in a geographical context

Hazards are the outcome of the interaction between natural environments and phenomena, on the one hand, and human environments, on the other. The hazards can be geophysical (volcanoes and earthquakes), atmospheric (storms and drought) or human-induced (some wildfires).

The way in which a hazard is perceived often depends upon the economic circumstances and cultural background of those experiencing the impacts of the hazard. Overall, hazards have social, economic and political impacts with both short- and long-term consequences.

A number of factors influence the nature of the human responses to a hazard. Possible responses are:

- fatalism
- prediction
- adjustment/adaptation
- mitigation
- management
- risk sharing

Clearly, responses are also linked to aspects of the hazard itself. These include the hazard's frequency, intensity, magnitude and distribution.

All of the above are linked to the level of development of the area in which the hazard occurs.

The Park model of human responses to a hazard sketches the phases following a hazard event, while the hazard management cycle takes into account the recurring nature of natural hazards.

Practice questions ?

1 Define the term 'natural hazard'.

....

....

2 Study Figure 5.1. Outline the activities that might take place during the relief phase.

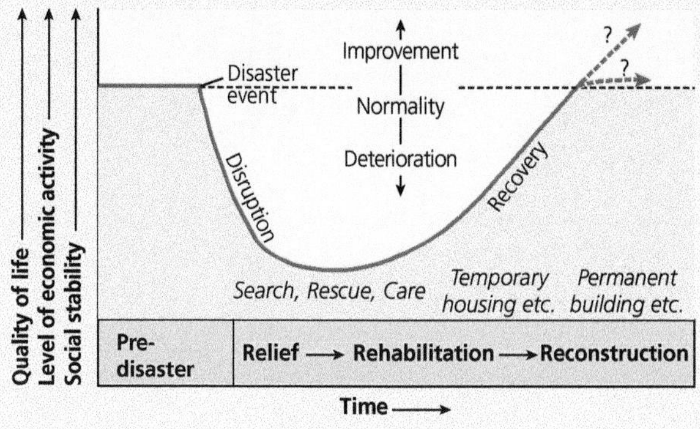

Figure 5.1 The Park model of human responses to a hazard

....

....

....

....

3   Complete Table 5.1.

Table 5.1

Human response to a hazard	Definition	Example
Fatalism		
Prediction		
Adjustment/adaptation		
Mitigation		
Management		
Risk sharing		

4   Using Figure 5.2 and examples from your studies, explain the extent to which the time interval between disasters impacts on an area's response to those disasters.

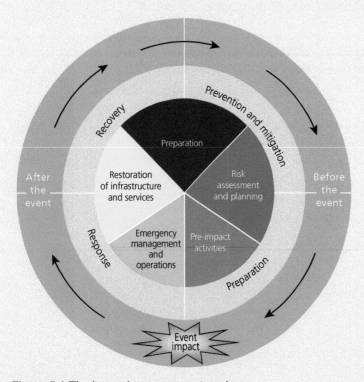

Figure 5.2 The hazard management cycle

....

....

....

....

....

Plate tectonics

Plate tectonics is the theory that the Earth's outer shell is divided into several rigid plates that glide over a rocky inner layer. These plates vary in shape and size and move relative to one another. There are nine major plates and several smaller ones.

Most of the boundaries between these plates are geologically active.

- Constructive plate boundaries occur where plates move away from each other and fresh magma wells up to create new crust (e.g. the Mid-Atlantic Ridge).

- Destructive plate boundaries are where crust sinks into the mantle and melts (e.g. the Pacific 'Ring of Fire').

- Conservative plate boundaries occur where crust is neither destroyed nor created as plates pass one another (e.g. the San Andreas Fault).

Localised heating at the core–mantle boundary causes a plume of magma to rise through the mantle and eat into the plate above at what is called a 'hot spot'. When this lava breaks through to the surface, active volcanoes form above the spot.

Practice questions ?

5 Describe the main features (location, temperature, chemistry etc.) of the following:

a The oceanic crust

....

....

....

....

....

....

b The continental crust

....

....

....

....

....

....

c The asthenosphere

....

....

....

....

    d   The mantle

....

....

....

....

    e   The core

....

....

....

....

6   Using Figure 5.3 and your own knowledge, outline how each of the two theories (ridge push and slab pull) explains how crustal plates move.

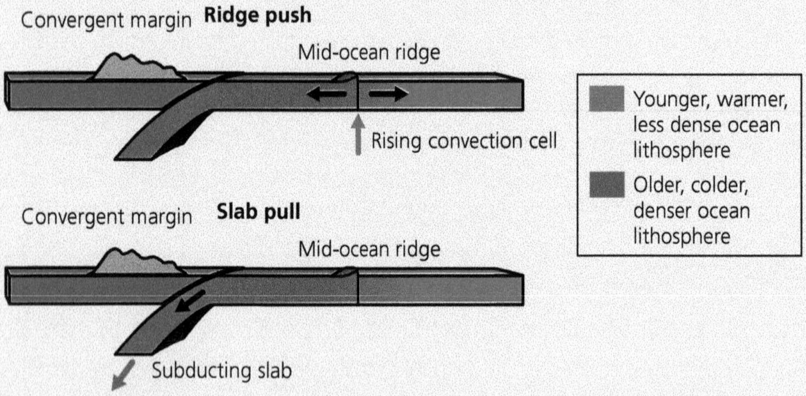

Figure 5.3 Ridge push and slab pull theories of crustal movement

Source: adapted from British Geological Survey

Worked example

One of the main differences between the theories is that they apply to different oceanic plate boundaries.

> **Application (AO2):** Shows understanding straight away. Shows that there is a comparison being made.

In the ridge push theory, molten magma that rises at a mid-ocean ridge is hot, heating the rocks around it. As the rocks at the ridge are heated, they expand and become elevated above the surrounding sea floor. This elevation produces a slope down and away from the ridge. As the newly formed rock ages and cools, it becomes denser. Gravity then causes this older, denser lithosphere to slide away from the ridge, down the sloping asthenosphere. As the older, denser lithosphere slides away, new molten magma wells up at the mid-ocean ridge, as shown in Figure 5.3, eventually becoming new lithosphere. It is thought that the cooling, subsiding rock exerts a force on spreading lithospheric plates, driving their movements.

> **Knowledge (AO1) and application (AO2):** Good knowledge and understanding of one side of the comparison.

The slab pull theory concentrates on subduction boundaries. One plate is denser and heavier than the other plate. The denser, heavier plate begins to subduct beneath the plate that is less dense. The edge of the subducting plate is much colder and heavier than the mantle, so it continues to sink, pulling the rest of the plate along with it. The force that the sinking edge of the plate exerts on the rest of the plate is called slab pull. Currently, many scientists consider slab pull to be a much stronger factor than ridge push in the movement of tectonic plates.

> Knowledge (AO1) and application (AO2): Good knowledge and understanding of the other side of the comparison.

7 What do you understand by the term 'sea floor spreading'?

....

....

....

8 To what extent does the mantle convection hypothesis explain plate movements at destructive boundaries and constructive plate margins?

....

....

....

....

....

9 Use plate tectonic theory to account for the evolution of oceans as shown in Figure 5.4.

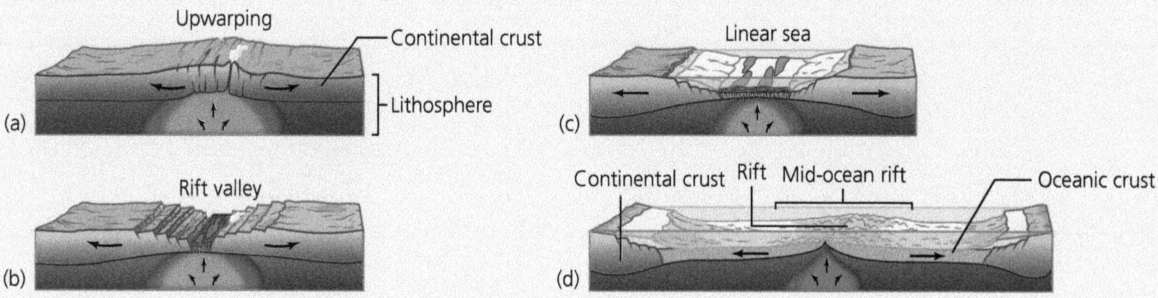

Figure 5.4 The evolution of a constructive plate margin

....

....

....

....

....

....

....

....

10 Explain why earthquakes found at constructive margins:

a have shallow foci

....

....

....

b cause little damage

....

....

11 The magma and lavas produced at constructive plate margins are low in silica content. To what extent does this affect the nature of the volcanic eruptions found at this type of boundary?

....

....

....

....

....

12 Study Figures 5.5 and 5.6. Account for the fact that earthquakes occur at these two types of boundary.

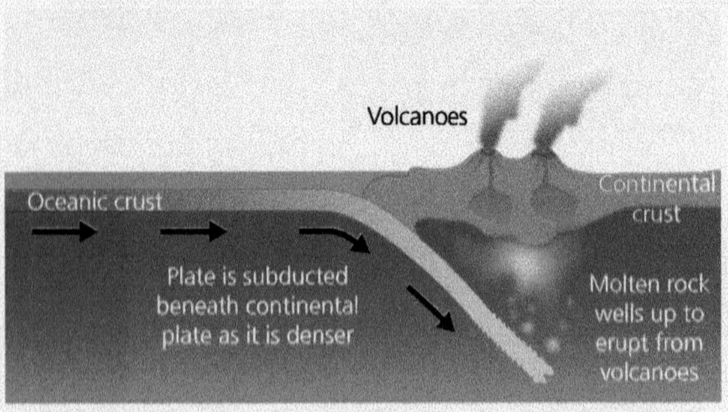

Figure 5.5 An ocean–continent destructive plate boundary

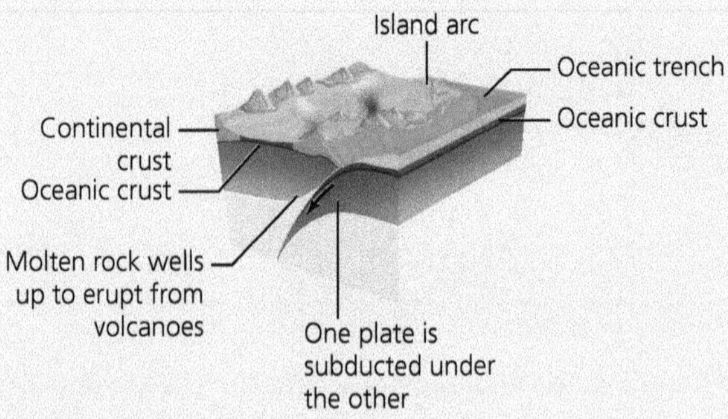

Figure 5.6 An ocean–ocean destructive plate boundary (an island arc)

....

....

....

....

13 Describe the nature of the volcanic eruptions at the two types of boundary shown in Figures 5.5 and 5.6.

....

....

....

....

....

....

....

....

14 Explain why collision boundaries such as that between the Indo-Australian plate and the Eurasian plate produce many earthquakes.

....

....

....

....

....

....

15 Explain why earthquakes at conservative plate boundaries are potentially so destructive.

....

....

....

....

....

....

16 Complete Table 5.2 with examples and details of tectonic landforms.

Table 5.2

Landform	Named example	Location	Description	Tectonic context
Ocean ridge				
Rift valley				
Young fold mountains				
Deep sea trench				
Island arc				

17 Using Figure 5.7, explain how the moving Pacific plate and the magma plume work together to form a chain of islands.

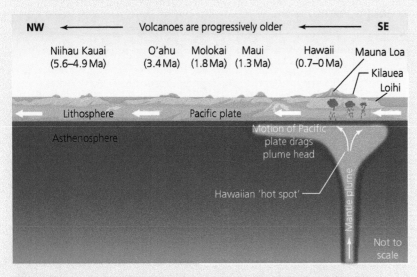

Figure 5.7 A chain of 'hot-spot' volcanoes in the Hawaiian Islands

....

....

....

....

....

....

....

....

Volcanic hazards

Volcanoes are mainly found along tectonic plate boundaries and above hot spots. Those found along constructive plate boundaries often have:

- low-viscosity basaltic lava that flows easily

- volcanic ash eruptions

Those found along destructive margins often have:

- more viscous acidic lava

- violent eruptions, with lots of steam, gas, and ash

Hot-spot volcanoes have very runny lava and form shield volcanoes.

A volcanic hazard refers to any potentially dangerous volcanic process which impacts on people and/or property or which negatively impacts the productive capacity or sustainability of a population. Such hazards include:

- nuées ardentes (pyroclastic flows)

- lava flows

- mudflows (lahars)

- volcanic ash and bombs (tephra)

- gas and acidic rainfall

Although it is impossible to prevent volcanic eruptions, it is possible to reduce their impact by proper management, both in preparation for and in response to volcanic risk.

Practice questions ?

18 Describe the distribution of volcanic activity as shown in Figure 5.8 and relate it to the theory of plate tectonics.

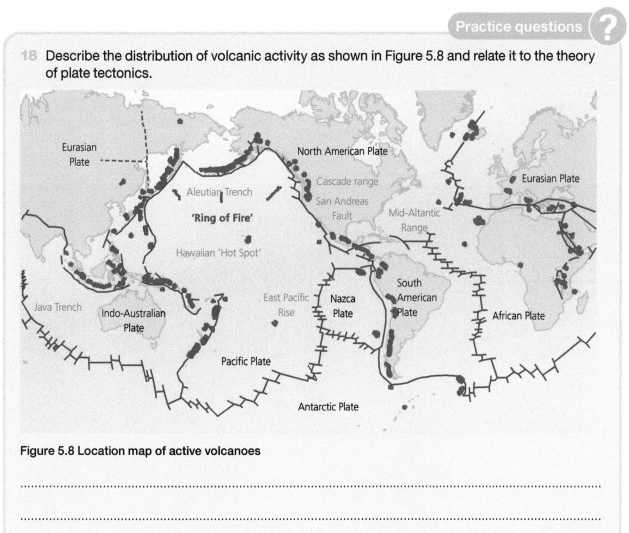

Figure 5.8 Location map of active volcanoes

....

....

....

....

....

....

19 Figure 5.9 shows a number of different volcanic hazards.

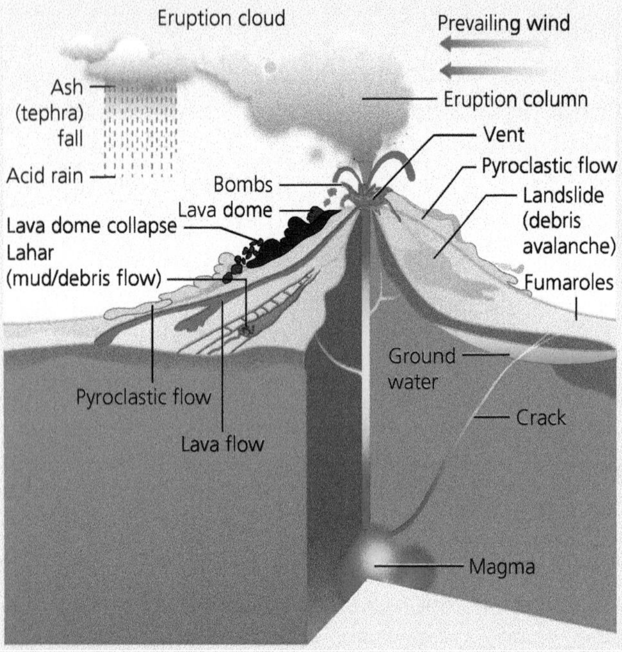

Figure 5.9 A simplified diagram showing some of the most common forms of volcanic hazard

For each of the following volcanic hazards (shown in the figure), name and locate an example, explain the cause(s) and describe the effects.

a   Nuées ardentes (pyroclastic flows)

....

....

....

....

....

b   Lava flows

....

....

....

....

....

c   Mudflows (lahars)

....

....

....

....

....

d   Pyroclastic and ash fallout (tephra)

....

....

....

....

....

e   Gases/acid rain

....

....

....

....

....

20 Outline **two** pieces of evidence that scientists use to predict forthcoming volcanic hazards.

....

....

....

....

....

21 What steps can be taken by individuals and organisations to reduce the impact of volcanic hazards?

....

....

....

....

22 For **one** recent hazardous volcanic event, do the following:

　a　Name and locate the event.

　..

　..

　b　Briefly describe the hazardous nature of the event.

　..

　..

　..

　..

　c　Outline the social, economic and environmental impacts of the event.

　..

　..

　..

　..

　..

　..

　d　Describe the human responses to the event.

　..

　..

　..

　..

　..

　..

Seismic hazards

Seismicity refers to the geographic and historical distribution of earthquakes. Seismic hazards occur when the physical results of an earthquake impact on people and/or property or negatively impact the productive capacity or sustainability of a population.

Almost all earthquakes occur within the crust in bands along:

- oceanic ridges and associated transform faults
- conservative plate margins
- broad zones below and behind destructive plate boundaries

Earthquakes are produced because the crust is constantly moving, causing a build-up of stress within rocks. When this pressure is released, shock waves travel out in all directions. Those that reach the surface radiate from the epicentre in an intense shaking motion.

When the earthquakes occur close to or under dry land, the surface waves damage human-made structures, which can cause people to be:

• crushed in a collapsing building

• drowned in a flood caused by a broken dam or levee

• buried under a landslide

• burned in a fire

Rocks with high water content can be liquefied and flow, causing mudflows or building collapse.

Sometimes when earthquakes occur under oceans, the shock creates fast-moving, long-wavelength sea waves that travel far from the epicentre. They pile up on coastlines, causing immense damage.

Although scientists know most of the global locations where earthquakes are likely to occur, they still have great difficulty in predicting when they may occur. This makes planning and preparation difficult.

Practice questions ?

23 Describe the distribution of seismic activity as shown in Figure 5.10 and relate it to the theory of plate tectonics.

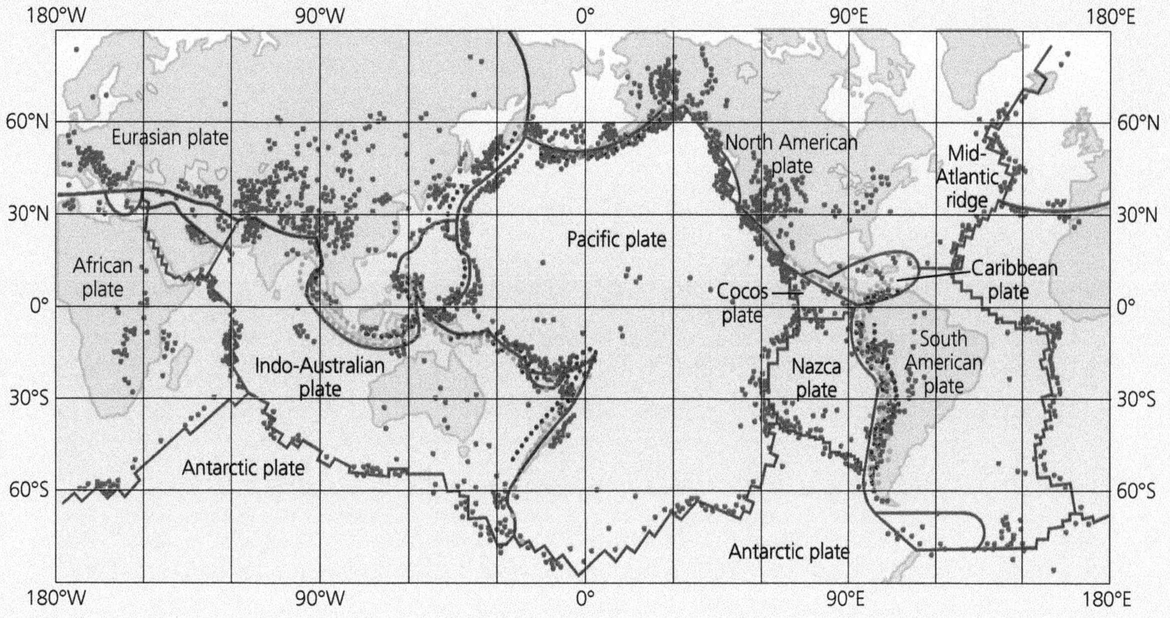

Figure 5.10 Earthquake locations for events between 1965 and 1995. The red dots are shallow earthquakes, the green are of intermediate depth, and the purple are deep

....

....

....

....

....

....

24 For each of the following seismic hazards, name and locate an example, explain the cause and describe the effects.

a Earthquakes/shock waves

....

....

....

....

....

b Liquefaction

....

....

....

....

....

c Landslide

....

....

....

....

....

d Tsunamis

....

....

....

....

....

25 Describe how the 'seismic gap theory' can help predict the location and timing of earthquakes.

....

....

....

....

....

....

26 What steps can be taken by individuals and organisations to reduce the impact of seismic hazards?

....

....

....

....

....

....

27 For **one** recent hazardous seismic event, do the following:

a   Name and locate the event.

....

....

b   Briefly describe the hazardous nature of the event.

....

....

....

c   Outline the social, economic and environmental impacts of the event.

....

....

....

....

....

d   Describe the human responses to the event.

....

....

....

....

....

....

Storm hazards

A tropical storm (or cyclone) is the generic term for a low-pressure system over tropical or subtropical waters, with organised convection (i.e. thunderstorm activity) and winds at low levels circulating either anticlockwise (in the northern hemisphere) or clockwise (in the southern hemisphere). The whole storm system may be 8,000 to 9,500 m high and 200 to 700 km wide, although it can sometimes be even bigger. It typically moves forward at speeds of $16\text{--}20\,\text{kmh}^{-1}$ but can travel as fast as $60\,\text{kmh}^{-1}$.

Tropical storms require a certain set of environmental conditions to develop. Once they have been triggered, they move westwards and northwards (southwards in the southern hemisphere) until they come under the influence of westerly winds. They then start to reverse their direction into an easterly path.

Tropical storms pose threats to coastal communities in the form of high winds, storm surges, coastal flooding and landslides. Their impact is huge. Between 1980 and 2009, 466 million people were affected, including 412,644 deaths and 290,654 injuries. The primary cause of cyclone-related mortality in both developed and less developed countries was storm surge drowning.

Practice questions ?

28 With the help of Figure 5.11, describe the structure of a typical tropical revolving storm.

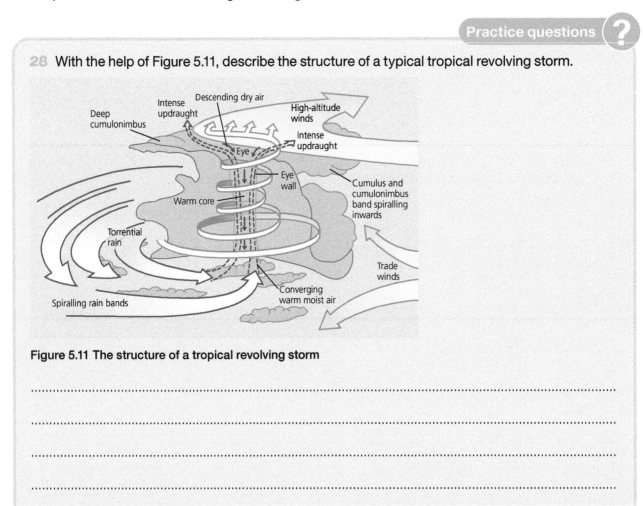

Figure 5.11 The structure of a tropical revolving storm

....

....

....

....

....

....

29 Describe the processes involved in the formation of a tropical storm.

....

....

....

....

....

....

....

30 Using Figure 5.12, describe and account for the global distribution of tropical storms and the time of their occurrence.

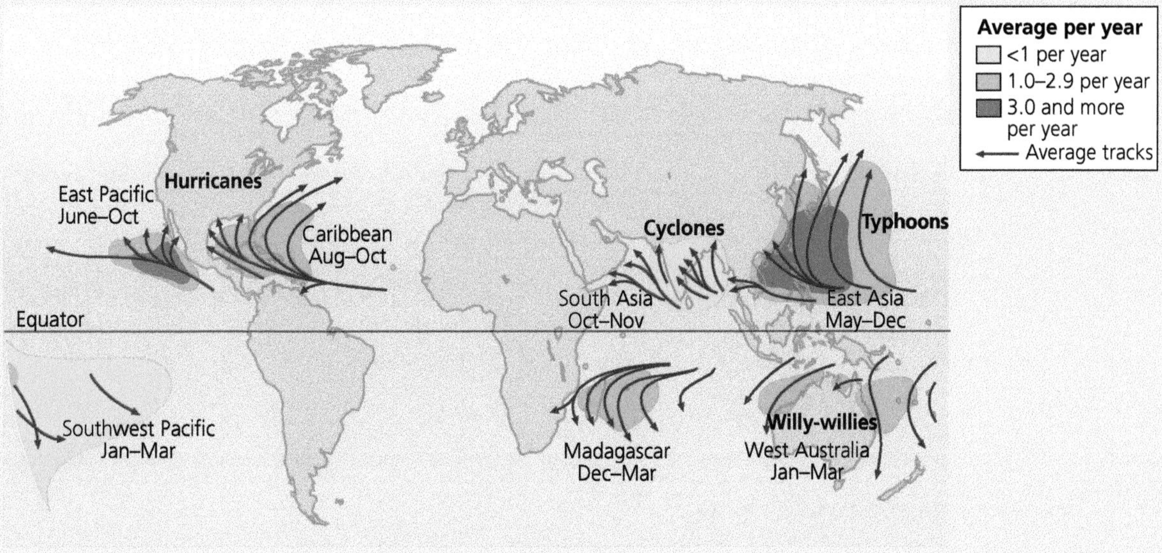

Figure 5.12 Global distribution and seasons of tropical storms

....

....

....

....

....

....

....

....

31 Outline **two** ways in which meteorologists can predict both the track and the intensity of a tropical storm.

32 For each of the following tropical storm hazards, name and locate an example, explain the cause and describe the effects.

a   High wind speeds

....

....

....

....

....

b   Storm surge and coastal flooding

....

....

....

....

....

c   Intense rainfall and river flooding

....

....

....

....

....

d   Landslides

....

....

....

....

....

33 What steps can be taken by organisations and individuals to reduce the impact of storm hazards?

....

....

....

....

....

....

34 Why might a decreased proportion of deaths and injuries be observed in the aftermath of tropical storms as a result of improved early warning systems and evacuation?

....

....

....

35 To what extent do the impacts of tropical storms and the human responses depend on both the physical nature of the environment and the level of economic development?

....

....

....

....

....

....

....

....

36 For **one** recent tropical storm, do the following:

 a   Name and locate the event.

....

....

 b   Briefly describe the hazardous nature of the storm.

....

....

....

....

 c   Outline the social, economic and environmental impacts of the storm.

....

....

....

....

....

....

 d   Describe the human responses to the event.

....

....

....

....

....

....

Fires in nature

A wildfire is an unplanned fire found in a natural area such as a forest, grassland or moorland. Such fires are often caused by human activity (abandoned barbecues, cigarette ends and so on) or natural lightning. The risk of wildfires increases in extremely dry conditions, such as drought, and during high winds.

The impacts of wildfires on the human environment include disruption or destruction of infrastructure (electricity, gas, water, telecommunications), housing, transport links (roads, railways) and agriculture.

Wildfires also destroy natural environments with a consequent loss of forest and other vegetation, as well as wild animal habitats.

Wildfires also impact weather and the climate by releasing large quantities of carbon and particulate matter into the atmosphere. This exacerbates climate change and can cause a range of health issues, including respiratory and cardiovascular problems.

Despite recent news coverage of wildfires in the USA and Australia, it is important to note that fatalities are relatively few and extensive wildfires have been around for millennia. Research has found a significant increase in burn severity in US forests and it is believed that outdated management practices (e.g. fire suppression) rather than climate change have led to major changes in the forests and resulting fires.

It is important to note that there are marked differences between wildfires and other natural hazards.

Practice questions ?

37 Complete Table 5.3.

Table 5.3 Conditions favouring wildfires

Factor	Impact on wildfire
Vegetation type	
Fuel characteristics	
Climate	
Recent weather conditions	
Fire behaviour	

38 Outline how wildfires can be started.

....

....

....

....

....

....

39 What factors need to be taken into account when producing a wildfire risk assessment strategy?

....

....

....

....

....

....

....

....

40 Explain how outdated fire management practices may be contributing to the severity of wildfire hazards.

....

....

....

....

....

....

41 For a recent wildfire event you have studied, describe the impacts of that fire in terms of the following:

a   Environmental impacts

....

....

....

....

b   Social impacts

....

....

....

....

c Economic impacts

....

....

....

....

d Political impacts

....

....

....

....

Exam-style questions

Exam-style set 1

1 Outline how risk management can reduce the impacts of wildfires. (AO1) **4 marks**

....

....

....

....

....

....

2 The United Nations Office for Disaster Risk Reduction has produced statistics regarding the number of natural disasters and the total costs of the damage incurred.

Table 5.4 shows the figures for the ten countries with the most disasters between 2005 and 2014.

Table 5.4

Country	Number of natural disasters	Rank	Total damage ($ billion)	Rank	d	d²
China	286	1	265	2	–1	1
USA	212	2	443	1	+1	1
Philippines	181	3	16	7	–4	16
India	167	4	47	4	0	0
Indonesia	141	5	11	8	–3	9
Vietnam	73	6	7	9	–3	9
Afghanistan	72	7	0.16	10		
Mexico	64	8	26	5	3	9
Japan	62	9	239	3		
Pakistan	59	10	25	6	4	16

Calculate the Spearman's rank correlation coefficient for the two sets of data by completing Table 5.4 and using the formula:

$$R_s = 1 - \left(\frac{6 \sum d^2}{n^3 - n} \right)$$

where:

R_s is the Spearman's rank correlation coefficient

n is the number of pairs of variables

$\sum d^2$ is the sum of the differences in rank squared.

Critical values for Spearman's rank where $n = 10$:

n	Significance level	
	0.05	0.01
10	±0.564	±0.746

Complete Table 5.4 and interpret your Spearman's rank result using the critical values above. (AO3)

7

6 marks

....

....

....

....

3 Using Table 5.5 and your own knowledge, assess the usefulness of the volcanic explosivity index (VEI) when considering the impacts of a volcanic eruption. (AO1, AO2)

10

9 marks

Table 5.5 The volcanic explosivity index

Volcanic explosivity index (VEI)	Eruption rate/ $kg\,s^{-1}$	Volume of ejecta/ m^3	Eruption column height/ km	Duration of continuous blasts/h	Troposphere/ stratosphere injection	Qualitative description	Example
0 Non-explosive	10^2–10^3	$<10^4$	0.8–1.5	<1	Negligible/ none	Effusive	Kilauea, erupts continuously
1 Small	10^3–10^4	10^4–10^6	1.5–2.8	<1	Minor/none	Gentle	Nyiragongo, 2002
2 Moderate	10^4–10^5	10^6–10^7	2.8–5.5	1–6	Moderate/ none	Explosive	Galeras, Colombia, 1993
3 Moderate–large	10^5–10^6	10^7–10^8	5.5–10.5	1–12	Great/ possible	Severe	Nevado del Ruiz, 1985
4 Large	10^6–10^7	10^8–10^9	10.5–17.0	1->12	Great/ definite	Violent	Mayon, 1895 Eyjafjallajökull, 2010
5 Very large	10^7–10^8	10^9–10^{10}	17.0–28.0	6->12	Great/ significant	Cataclysmic	Vesuvius, AD 79 Mt St Helens, 1980
6 Very large	10^8–10^9	10^{10}–10^{11}	28.0–47.0	>12	Great/ significant	Paroxysmal	Mt Pinatubo, 1991
7 Very large	$>10^9$	10^{11}–10^{12}	>47.0	>12	Great/ significant	Colossal	Tambora, 1815
8 Very large	–	$>10^{12}$	–	>12	Great/ significant	Terrific	Yellowstone, millions of years ago

....

....

....

....

....

....

....

....

....

....

4 With reference to **one or more** seismic events that you have studied, assess the extent to which the hazards created can be managed effectively. (AO1, AO2) **9 marks** ⏱ 10

....

....

....

....

....

....

....

....

....

....

....

....

5 Using a named example of a place that has suffered a recent hazardous event, assess the extent to which people's lived experience of that place has been affected by the hazardous event. (AO1, AO2) **20 marks** ⏱ 25

Plan and write your answer on a separate sheet of paper and keep it with your workbook.

Exam-style set 2

1 Describe **two** forms of storm hazard. (AO1) **4 marks** ⏱ 5

....

....

....

....

....

....

....

....

2 Analyse the data shown in Figure 5.13. (AO3)   6 marks

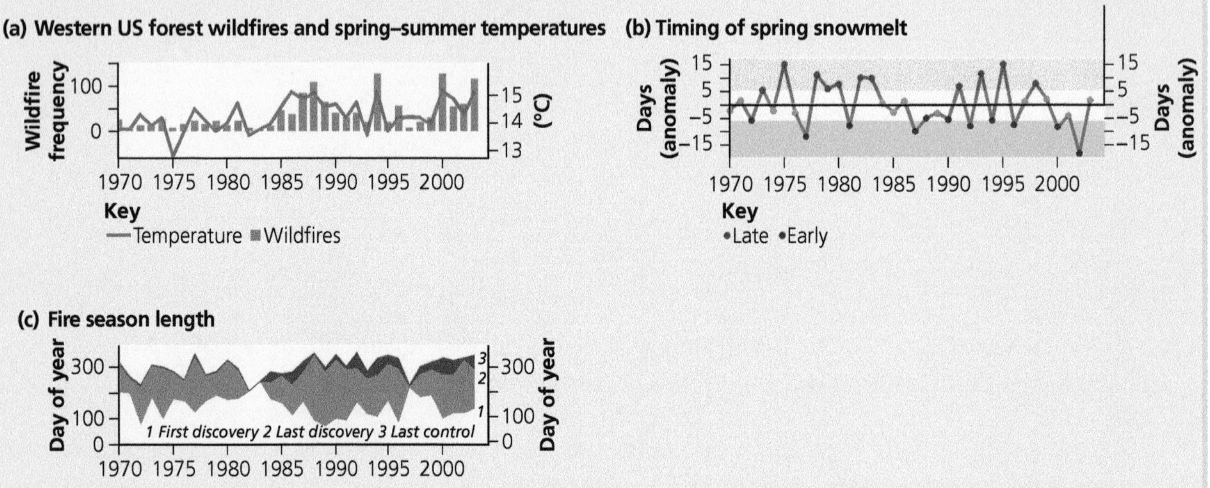

(a) Western US forest wildfires and spring–summer temperatures   **(b)** Timing of spring snowmelt

Key
——Temperature ■Wildfires

Key
●Late ●Early

(c) Fire season length

1 First discovery 2 Last discovery 3 Last control

Figure 5.13 Annual frequency of large (> 400 ha) western US forest wildfires (bars) and mean March–August temperatures (a); timing of spring snowmelt for western USA (b); and fire season length (c)

....

....

....

....

....

....

....

....

....

3 Using Figure 5.14 and your own knowledge, assess the extent to which destructive plate boundaries could be viewed as very susceptible to earthquake hazards. (AO1, AO2)

10

9 marks

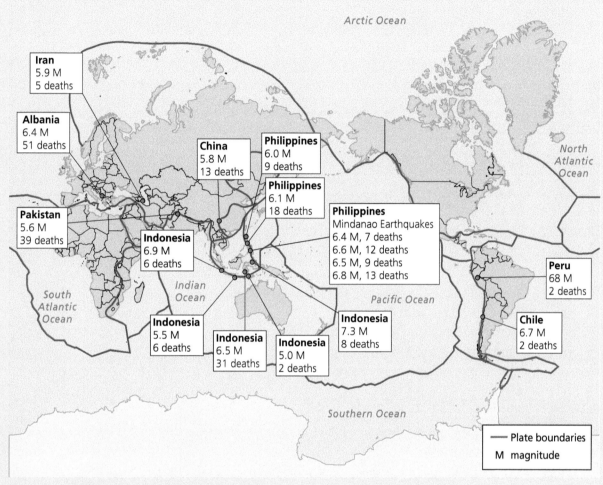

Figure 5.14 Global overview of 2019 earthquakes, showing events with a magnitude of 5.0 M or greater and more than one fatality

Source: adapted from Emergency Response Coordination Centre, European Commission

....

....

....

....

....

....

....

....

....

....

4   Assess the extent to which the positions of young fold mountains, rift valleys, ocean ridges, volcanoes, deep sea trenches and island arcs support the theory of plate tectonics. (AO1, AO2)

10

9 marks

....

....

....

....

....

....

....

....

....

....

5   Using an example of a multi-hazardous environment beyond the UK, assess the extent to which human qualities and responses (such as fatalism, adaptation, mitigation and management) contribute to its continuing human occupation. (AO1, AO2)

25

20 marks

Plan and write your answer on a separate sheet of paper and keep it with your workbook.

Additional extended prose questions

Plan and write your answers on separate sheets of paper and keep them with your workbook.

1   'The economic, social and political character of a local community reflects the presence and impacts of hazard risk and the community's responses to that hazard.'

25

For a place at a local scale that you have studied, assess the extent to which you support this view. (AO1) (AO2)

20 marks

2   To what extent do the human responses to a hazard/s depend on the character of the place affected by that hazard/s? (AO1) (AO2)

25

20 marks